Springer Theses

Recognizing Outstanding Ph.D. Research

Aims and Scope

The series "Springer Theses" brings together a selection of the very best Ph.D. theses from around the world and across the physical sciences. Nominated and endorsed by two recognized specialists, each published volume has been selected for its scientific excellence and the high impact of its contents for the pertinent field of research. For greater accessibility to non-specialists, the published versions include an extended introduction, as well as a foreword by the student's supervisor explaining the special relevance of the work for the field. As a whole, the series will provide a valuable resource both for newcomers to the research fields described, and for other scientists seeking detailed background information on special questions. Finally, it provides an accredited documentation of the valuable contributions made by today's younger generation of scientists.

Theses may be nominated for publication in this series by heads of department at internationally leading universities or institutes and should fulfill all of the following criteria

- They must be written in good English.
- The topic should fall within the confines of Chemistry, Physics, Earth Sciences, Engineering and related interdisciplinary fields such as Materials, Nanoscience, Chemical Engineering, Complex Systems and Biophysics.
- The work reported in the thesis must represent a significant scientific advance.
- If the thesis includes previously published material, permission to reproduce this must be gained from the respective copyright holder (a maximum 30% of the thesis should be a verbatim reproduction from the author's previous publications).
- They must have been examined and passed during the 12 months prior to nomination.
- Each thesis should include a foreword by the supervisor outlining the significance of its content.
- The theses should have a clearly defined structure including an introduction accessible to new PhD students and scientists not expert in the relevant field.

Indexed by zbMATH.

Simone Gargiulo

Electromagnetic Processes of Nuclear Excitation

From Direct Photoabsorption to Free Electron and Muon Capture

Doctoral Thesis accepted by
École Polytechnique Fédérale de Lausanne, Lausanne,
Switzerland

Springer

Author
Dr. Simone Gargiulo
Institute of Physics (IPhys)
École Polytechnique Fédérale de Lausanne (EPFL)
Lausanne, Switzerland

Supervisor
Prof. Fabrizio Carbone
Laboratory for Ultrafast Microscopy and Electron Scattering
Institute of Physics (IPhys)
École Polytechnique Fédérale de Lausanne (EPFL)
Lausanne, Switzerland

ISSN 2190-5053 ISSN 2190-5061 (electronic)
Springer Theses
ISBN 978-3-031-56261-7 ISBN 978-3-031-56262-4 (eBook)
https://doi.org/10.1007/978-3-031-56262-4

This Springer imprint is published by the registered company Springer Nature Switzerland AG
The registered company address is: Gewerbestrasse 11, 6330 Cham, Switzerland

Paper in this product is recyclable.

Sometimes it is the people no one can imagine anything of who do the things no one can imagine.

—Alan Turing

To my parents, whose endless efforts and altruism have made my accomplishments possible, and to all those who have shaped who I am today.
I am deeply grateful for your encouragement, support, and patience as I followed my dream of becoming a scientist.

Supervisor's Foreword

Recent technological advances allowed for the manipulation of material's properties in out-of-equilibrium conditions. In this thesis, the possibility of applying these methods to control nuclear phenomena is addressed both theoretically and experimentally. When a free electron is captured by an ion having a vacancy in one of its deep core-levels, if the energy of the electron plus the binding energy of the capturing orbital matches one of the atom's nuclear excitations, an isomeric transition can be induced. In this thesis, it has been shown theoretically that if the capturing ion's electronic structure is out of equilibrium, the cross section of such effect can be enhanced. In a separate work, it has also been shown that it is possible to further enhance such a cross section if the wavefunction of the free electron is engineered prior to the capture to better match the energy-momentum conservation of a specific orbital. Building on these ideas, a new effect is also proposed by exploiting the possibility for any neutral ions to capture muons in one of their naturally empty muonic orbitals, potentially yielding nuclear excitations at much higher energies, in the order of MeV. These results offer a new perspective on the possibility to use modern tools to control the nuclear properties of matter. To implement these ideas, a new instrument has been developed and described in this thesis, aiming at measuring the fluorescence of excited nuclei upon a pulsed excitation. Currently, it is commonly assumed that a laser-driven plasma can be used to generate X rays to directly photoexcite a nuclear transition. This thesis reports an extensive experimental campaign aiming at reproducing and benchmarking this effect, and shows that the process reported in the literature cannot be ascribed to a nuclear transition. Furthermore, an extensive modeling of the data is provided, offering a quantitative perspective of the excitation probabilities of nuclei in a laser-driven plasma scenario. Overall, this thesis provides a historically accurate description of the electromagnetic excitation mechanisms of isomers and reports theoretical and experimental original research aimed at discussing the feasibility and perspective of their active manipulation. These results have implications for both the discovery of novel methods for nuclear energy harvesting and storage, and for fundamental research in nuclear physics.

Lausanne, Switzerland
May 2023

Fabrizio Carbone

Abstract

In the vast expanse of the Universe and on our planet, nuclei exist in a state of excitation. These nuclei, known as nuclear isomers, possess unique properties that make them play a crucial role in diverse domains of physics. In nucleosynthesis, the process by which new atomic nuclei are formed inside stars, isomers can influence the rates of nuclear reactions, leading to variations in the production and abundance of elements we see around us. Additionally, some isomers hold potential applications in nuclear technologies, including the energy sector. These excited nuclear states can persist for varying periods, from fractions of a second to billions of years and beyond, before decaying to their ground state. If harnessed, feeding and depleting these isomers could represent a clean and high-density way to store and release energy on demand. The quest for efficient dynamical population control of nuclear isomer has long captivated the imagination of physicists, yet this elusive goal remains beyond our grasp.

In this dissertation, I examine the potential of employing nuclear excitation mechanisms as viable tools for achieving such manipulation. Several processes of nuclear excitations from both theoretical and experimental perspectives are explored: direct photoabsorption, nuclear excitation by electron capture (NEEC), and nuclear excitation by free muon capture (NEμC).

This thesis begins by delving into the historical framework of nuclear excitation by electron capture (NEEC), a process that was proposed in 1976 and is yet to be comprehended. A recently claimed observation has sparked new interest in nuclear excitation processes as a way to release the energy trapped in isomers. However, the irreconcilability between the first observation, the theoretical framework, and the recent repetition of the experiment reveals that much remains to be learned.

Regardless of the specific process being examined, the primary goal is to increase the likelihood of their occurrence. One such possibility involves NEEC taking place in excited ions, where the screening effect of other electrons provides nearly resonant orbitals where capture can occur. This process was initially proposed to mitigate the discrepancy between the experimental finding and the theoretical prediction. In this new setting, three orders of magnitude increase in the NEEC cross sections for ^{73}Ge is found theoretically. Another approach enabling the manipulation of the NEEC cross section involves engineering the electron wavefunction that undergoes capture. This technique not only demonstrates an increased occurrence of NEEC but also highlights the potential to alter the shell where the highest capture takes place.

The second mechanism, NEμC, occurs in exotic muonic atoms. The process is introduced as a counterpart to NEEC, with the electron being replaced by a muon. It follows a presentation of the framework within which this process has emerged and how it changes the paradigm in comparison to NEEC. Owing to the increased proximity of muons to the nucleus, this process has been found to exhibit cross sections that are several orders of magnitude higher than NEEC for excitations in the MeV range. By examining the unique properties of NEμC, insights into the process and its potential applications are provided, including muon-induced fission.

Lastly, nuclear excitations are studied in the context of a laser-generated plasma scenario, where nuclei might be excited through the resonant absorption of a photon, together with other competing processes (such as inelastic electron scattering). The design and implementation of a tabletop setup for generating keV-hot plasma upon femtosecond laser irradiation are presented. The experimental work has been conducted on a ^{181}Ta target using a time-dependent X-ray spectroscopic technique. The absence of a clear decay signal raises the question of whether the excitation of the ^{181m}Ta isomer has ever been observed in this context.

This dissertation aims to deepen the understanding of nuclear excitation mechanisms, emphasizing their complexities and potential for both further fundamental research and practical applications. Throughout the chapters, various ideas are mentioned for future research that could expand the scope of the physics here discussed.

Acknowledgements

To create freedom for oneself and also a sacred No to duty: for that, my brothers, the lion is required. To take the right to new values—that is the most terrible taking for a carrying and reverent spirit. [...] The child is innocence and forgetting, a new beginning, a game, a wheel rolling out of itself, a first movement, a sacred yes-saying.

Three metamorphoses of the spirit I named for you: how the spirit became a camel, and the camel a lion, and finally the lion a child.

—Friedrich Nietzsche, Thus Spoke Zarathustra

This has been a fantastic journey! Let's be clear, it has been challenging and full of uncertainties, and the contentment does not certainly come from having completed it. I am grateful as I have learned a bit more about myself and the world that surrounds me—hopefully undergone a metamorphosis of the spirit. Nonetheless, the path toward the spirit of a child—taking the right for oneself to create new values—is only at the beginning.

I slowly realized, over the years, how vital the care and attention my parents had during my childhood was. The certainty of always having them by my side has been the anchor and security that allowed me to devote myself to the pursuit of my passions. To them goes my greatest gratitude; for the years they spent chasing after me and making me a better person. I thank them for the warm, bonded family they have given to me and my siblings. I thank my sister Sara and my brother Costantino. Growing up together has been surely one of the best parts of my life, which I look back on with nostalgia. Thanks for your patience and comprehension.

I have been gifted to have met you, Mary Joy; as I can behave as Ivan Karamazov only because you are my Alëša. Gentleness, care, love, and *joy* are the characteristics that find immense richness in you. Thank you for being the exceptional person you are! Thanks for the way you understand me and make me feel accepted. Thank you for being my strongest supporter under all circumstances. Thank you for showing me, through your kind example, how to improve in the various traits that are lacking in me. These years would not have been the same without you. "If I am really able to care for the sticky little leaves I shall only love them, remembering you". *Te voglj ben overamente.*

I am thankful to my grandmother Rosa and those who are no longer here: nonno Costantino and nonna Pina. Thank you for your teachings and your love. I know you are proud of me. I keep the memories of my wonderful childhood spent with you as the most precious baggage.

These years have been emotionally challenging for several reasons, and now I can't help but think of you Josef. You would surely have made some jokes today. I am thankful to have met you, but unfortunately, not for long enough. I thank Enzo, Antonio, and Ciccio, who have always been present over the years, even if just through a phone call. I also want to thank the people I met during my previous CERN experience: Daniele, Luca, Leanne, and Francesca. Although we have lost our daily routine together, it feels as if all this time has never passed.

I am grateful to have met Fabrizio Carbone along my academic and life path, as this has been a pivotal point that changed everything. He gave me the incredible opportunity to become a scientist (and a physicist) and realize my potential. He was not only a scientific PI for me but a person with whom I could openly discuss any concern about life. I thank him for his tireless support, the freedom he gave me, extensive and stimulating discussions, and for believing in me more than I do. Thank you for the atmosphere you created in the group, which allowed me to be creative and meet amazing people. Thank you for giving me so much!

I am thankful to Ross Koningstein at Google Inc. for enthusiastically supporting and funding this research project. All this would not have happened without his vision.

I have also been blessed to have found someone like Ivan Madan on this journey. I am grateful and indebted to you for the endless conversations and discussions. Thank you for your untiring perseverance in advising me on the best path to ultimately "become who I am". The time spent together has been precious. Thank you for setting up a creative environment in which our ideas could flourish. I will keep with me the indelible memory of the day I won an outdoor chess game against you in downtown Lausanne. Thank you for being your *typical* self and the outstanding person you really are. Thank you for all you have taught me as a scientist and as a person; I owe you a lot. I hope that 30 years from now, we can still share these memories while drinking a pleasant coffee in Naples, discussing how our spirits have finally evolved and how we set a new beginning.

I am thankful for having met you, Siham (*Sisi*). You have often lightened my days simply with a hug or a laugh. You have always been available in these 4 years (do you remember that day in the hospital with Paolo?), and I want to thank you warmly. Spending all these years together was a lot of fun. I have missed you very much since you left EPFL. I realize only now how much my daily routine and joy were also connected to your presence.

I want to thank Francesco for all the nice time we spent together, for the discussions, and for pushing me to try hiking. I still remember your arguments against the Swiss website that listed the route as being of the highest difficulty while you claimed it was entry-level. When I think of you, I associate you irretrievably with Barberousse or the bicycle ride home on a winter night and, of course, with food.

I also want to thank Gianmaria. Even if briefly, it was nice to share time together. Your daily 8 p.m. call before dinner has been a source of entertainment for years. These phone calls reminded me of the typical phone call between Assunta and Salvatore that all Neapolitans know. It has been essential to have you along this path.

I will be forever grateful to my beautiful "Italian office": Paolo, Francesco, Veronica, and Benoît (clearly an Italian name). Meeting all of us together was almost a planetary alignment. I doubt that I will ever have such a fun and cohesive working environment in the future. It was an excellent time spent together. I'll need pages to even start to share a single episode about every one of you: *Cooosa? WoW! Let's Barantize the data! Cra Cra! Assurdo! Lausanne is like San Francisco! Who is going to reimburse my time?* Paolo, you have entertained me so much over the years; I can't even imagine what it would have been like without you in the office (yes, maybe a little quieter). You and Siham have been my perfect accomplices in pranks and jokes, brightening these years with colors. Thank you, Ben, for your constant availability, help, and joy. Lately, our office has been further enriched by a newcomer from Bergamo (and now I know you are reading this with his voice). This marked a new point of reference for enjoyment but, above all, for stimulating discussions. I want to thank Hui-Yuan for simply being himself and for the way he was always welcoming with a smile or a hug. I will never forget your ability to stand in line for literally any activity during the conferences. I want to thank also Lukas, Alexey, Rémi, and Tom for the nice time we spent together and for our interesting discussions.

I want to thank also all the scientists with whom I had the opportunity to collaborate, especially Javier García de Abajo, Ido Kaminer, Adriana Pállfy, Yuanbin Wu, Vincenzo Grillo, and Eduardo Dias.

I want to warmly thank the committee for their extensive feedback on this work and the stimulating discussions during the defense. It was a memorable day that I will remember with affection. I would like to express my gratitude to Laura Elisa Marcucci for her support in publishing this thesis, and for all her kind words and encouragement. I also want to thank Chris Chiara and Jeff Carroll for the countless discussions and for having shared their knowledge on the field and on the NEEC process since day one. I want to thank them for the level of detail in their comments and explanations and the long emails of correspondence. All of this has been as a Rosetta Stone that helped me in the comprehension of the nuclear processes when I was learning to crawl in my research activity.

I am grateful to have met Amedeo Capozzoli at the University of Naples because, after all, he was the one who inflamed my critical thinking during my studies. Meeting him definitely changed my academic path, and I have held over these years, with affection, his priceless pearls of wisdom. I would like also to extend my gratitude to Claudio Curcio and Giuseppe D'Elia. Attending their classes was an important moment in my education, and I remember that period with great fondness.

Grazie a voi tutti. Grazie per ciò che siete stati e la ricchezza che mi avete donato.

Post Tenebras LUMES. From *left* to *right*: Ivan Madan, Phoebe Tengdin, Le Yu, Paolo Usai, Francesco Barantani, Benoit Truc, Alexey Sapozhnik, Fabrizio Carbone, Leonor Camporeale, Tom LaGrange, Siham Benhabib, Veronica Leccese, Michele Caldara, Simone Gargiulo

Lausanne, Switzerland
October 2023

Simone Gargiulo

Contents

Chapter 1
Introduction

> *There is only one thing that I know for certain: the value judgments of human beings are undoubtedly guided by their desire for happiness and thus amount to an attempt to back up their illusions with arguments.*
>
> — Sigmund Freud, *Civilization and Its Discontents*

From the beginning of human civilization, the mastery of the resources around us has been essential for us to climb the evolutionary ladder. Taming the fire, intriguingly seen as a cultural triumph for foregoing the satisfaction of a drive [1], was undeniably transformative. With the simple act of gathering around hearths, we learned the value of group cooperation toward survival. This fundamental realization, together with the advent of language, has shaped the foundation of our system of values, social structures, along with our search for meaning [2].

To this day, we continue to fight to uncover the underlying laws that would enable us to comprehend our own nature. At the same time, we have managed to transcend this struggle, distilling—as revelations between dreams and consciousness—the attributes that have proven advantageous to our *progress*.[1] The magnitude of such accomplishments does not need to be mentioned. Reflecting on this, Freud wrote: «*What man, through his science and technology, has produced in this world, where he first appeared as a frail animal organism and where every individual of his species must still make his entry as a helpless babe - 'oh inch of nature!' - all this not only sounds like a fairy tale, but actually fulfills all - no, most - fairy-tale wishes.*» [1].

When I think back to the start of this research project and the journey through my Ph.D., I realize that it also felt much like embarking on a fairy tale story, or

[1] I know there should be an agreement here on the definition of what can be considered progress, but I rely on common perception.

S. Gargiulo, *Electromagnetic Processes of Nuclear Excitation*, Springer Theses,
https://doi.org/10.1007/978-3-031-56262-4_1

at least aiming to bring to life most of the *fairy-tale wishes*. During that period, a paper appeared in Nature [3] announcing the observation of a phenomenon predicted over four decades earlier. This process, called nuclear excitation by electron capture and abbreviated with NEEC, revived interest in the potential to manipulate nuclear transitions using external controls. The measured probability of this process, an astonishing 1%, pointed towards prospective applications. Resurfacing from history, the vision was to develop next-gen nuclear batteries [4] by storing energy in long-lived nuclear excitations, known as isomers.

The core idea is identical to the way radioisotopes are used for analogous purposes, where energy is released through beta or alpha decay from an unstable isotope. Yet, isomers distinguish themselves by possessing not only the typical decay channels available to the ground states but also the capability to undergo electromagnetic decays—either through internal conversion (resulting in an orbital electron being ejected) or by gamma-ray emission. This feature, in theory, allows for the storage and release of energy without altering the nature of the isotope used. Essentially, if we begin with an isotope that has a stable ground state (i.e., non-radioactive), following the procedures of energy storage and release, we should be able to return to that original stable state. In abstract terms, this cycle ensures that the isotope can be reused without turning it into a radioactive element. Moreover, unlike the passive approach generally employed with radioisotope batteries, where one generally waits for their natural decay (though even here one can speed up the decay up to some extent), the modus operandi with isomer wishes to be more active. The aim is to accelerate the decay by using nuclear excitation processes, pushing the isomer toward quickly decaying upper-lying levels (thus having a much shorter lifetime than the isomer itself), in order to subsequently release the stored energy. If one imagines a tiny amount of energy, few MeV, encapsulated within the excitation of every single nucleus that forms regular matter—which can be thought of as the fundamental building blocks of a nuclear battery—isomers possess a significant energy density compared to alternative sources [5, 6], as highlighted in Fig. 1.1. To help visualization, a building realized of ^{178}Hf measuring 40 m on each side could eventually store the world's entire energy requirement for a year, equivalent to 160 PW h.

Besides their relatively high energy density, isomers, even though they are formally metastable states, can possess lifetimes that vary enormously, spanning from a few nanoseconds to billions of years and beyond. As depicted in Fig. 1.2, the isomer with the longest known lifetime, $^{180\text{m}}$Ta, holds its excitation for a duration that exceeds even the accepted age of the Universe.

Given these traits, it is not surprising that isomers have been identified as compelling candidates for exceptionally compact long-term energy storage applications. Looking ahead, the implementation of such technology could potentially drive substantial advancements in both green energy transition and space explorations.

However, a pressing question emerges: how can we control the insertion or extraction of this energy? And here, my query is not about the conversion of nuclear decay into something functionally useful, but rather about the mechanisms through which this energy can be deliberately released or stored by nuclei in response to external stimuli. Historically, several nuclear excitation processes have been put forward

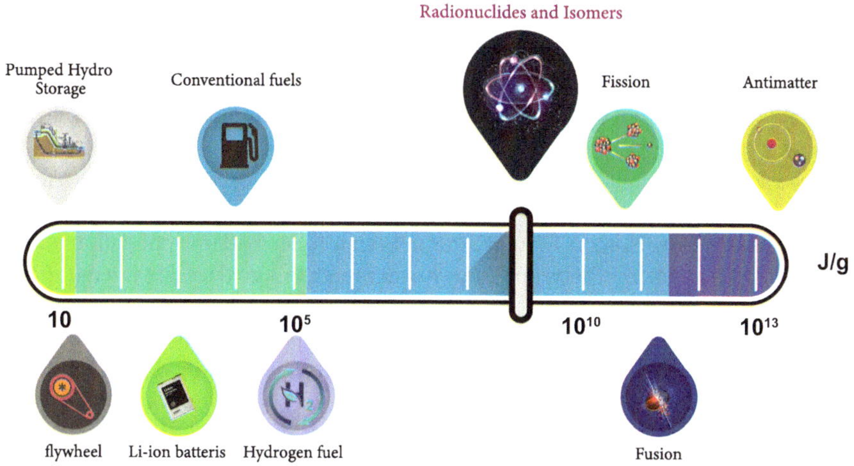

Fig. 1.1 Energy density in J/g for various sources. Image created in Adobe Illustrator using vector elements from Adobe Stock

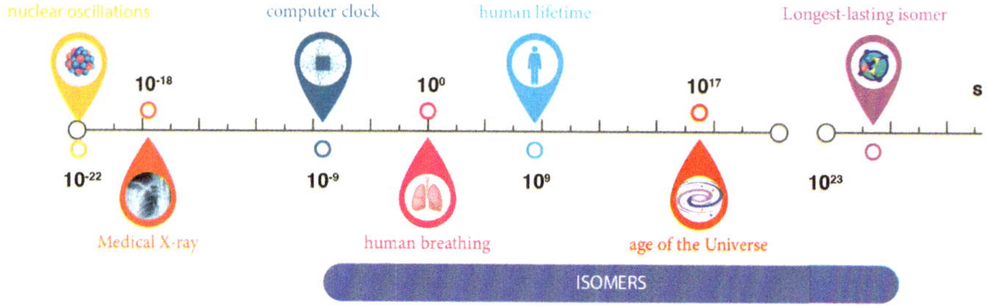

Fig. 1.2 Isomer lifetimes range extends from nanosecond to billion of years and beyond. The timescale is expressed in seconds. Image created in Adobe Illustrator using vector elements from Adobe Stock

as possible means of driving the isomer toward more unstable levels (referred to as gateways), thereby facilitating the decay. Despite substantial interest from the theoretical physics point of view and a plethora of proposals, not much progress has been achieved in practical terms. This refers not only to the realization of a real-world application but often even to the basic observation of such phenomena. Various elements might contribute to this: the low occurrence of these processes, the stringent conditions required, the necessity of large facilities for the experiments, and the challenges of each unique scenario. Furthermore, the simultaneous cooperation of multiple processes during nuclear excitation can make it challenging to isolate specific contributions. Consequently, many of these processes remain hidden today, without even delving into the potential of manipulating them!

If I were to summarize the single unifying theme across this thesis, it would be the endeavor to discover innovative methods to enhance nuclear excitation processes, ideally under external control. However, this is not an endpoint. It has served as the driving force propelling the ideas put forth in this work. Delving deep into the specific reasons for the low probability of detecting these phenomena has made several intuitions emerge. These intuitions then took shape in various ways, such as excited configurations, vortices, and muons, even though their underlying genesis was the same. Although in the upcoming chapters, we will explore several strategies to enhance the nuclear excitation processes, from our present understanding, the dream of harnessing isomers for real externally-controlled battery applications seems nevertheless distant, or to draw an analogy with Freud, this goal remains an unfulfilled *fairy-tale wish.*

In writing this thesis, my goal was to portray concepts not as an aseptic collection of knowledge, but as instruments that have assisted me in asking deeper and more relevant questions or simply ignited curiosity. Hence, when relevant, I have also included the thought process that led to the processes discussed. As I have learned, the framework and tools that shape our thinking are often more valuable than the individual thoughts themselves. I acknowledge that I may not have completely fulfilled my desire, but the process has been an incredibly valuable exercise.

In **Chap.** 2, I will present nuclear isomers starting from their discovery that happened in 1921. We will see in detail the properties of isomers, understand the reasons behind their emergence, and discuss the proposed strategies for their depletion. In this chapter, I followed the overall structure common to many topical reviews and to the book titled "Nuclear Isomers: A Primer" [7]. While doing so, I tried to condense all the concepts learned from several books that have been instrumental in this journey. With the hope of having written a harmonious and organic text, this may serve as a starting point for various branching paths that can be explored through the cited literature.

In the subsequent chapters, I discuss three mechanisms of nuclear excitation that might be used to feed or deplete isomers. These mechanisms include the capture of a free particle (electron or muon) or the absorption of a photon. In **Chap.** 3, I present an in-depth historical overview of the nuclear excitation by electron capture (NEEC) process. Within this historical framework, I will contextualize and discuss our contributions and findings. In **Chap.** 4, the process of nuclear excitation by free muon capture (NEμC) is introduced as the muonic counterpart of the NEEC process. In **Chap.** 5, it will be discussed the experiment I had the opportunity to design and implement at EPFL. This experiment is meant to study nuclear excitations in neutral plasma generated by a femtosecond laser pulse. An additional objective of this experiment was to investigate the feasibility of conducting nuclear physics experiments in a tabletop setting. This chapter contains a theoretical analysis of the experimental conditions, alongside estimates for both the direct photoexcitation and the nuclear excitation by inelastic scattering processes within a plasma environment. The presented experimental findings seem to suggest that nuclear excitation of the low-lying ^{181m}Ta isomer in a laser-generated plasma might not have been accomplished and observed, contrary to previous literature claims.

Finally, **Chap.** 6 contains some final thoughts and remarks, which could probably be called epistemological.

References

1. Freud S (1929) Civilization and its discontents
2. Peterson JB (2002) Maps of meaning: the architecture of belief. Routledge
3. Chiara C et al (2018) Isomer depletion as experimental evidence of nuclear excitation by electron capture. Nature 554:216–218
4. Aprahamian A, Sun Y (2005) Long live isomer research. Nat Phys 1:81–82
5. Litz M, Merkel G (2004) Control led extraction of energy from nuclear isomers. In: Proceedings for the army science conference (24th) 7
6. Carroll JJ (2018) Nuclear metastables for energy and power: status and challenges. Innovat Army Energy Power Mater Technol 36:289
7. Jain AK, Maheshwari B, Goel A (2021) Nuclear isomers: a primer. Springer Nature

Chapter 2
Nuclear Isomers

The aim is lacking; 'why?' finds no answer.

— Friedrich Nietzsche

It has been nearly 100 years since Otto Hahn, while searching for the precursor of actinium, realized that the newly discovered radioactive product UZ, was not only isotopic with the previously identified UX_2, but also shared with it the same mass number [1–4].

UX_2 and UZ, later identified with ^{234m}Pa and ^{234}Pa as shown in Fig. 2.1, represent the first example of isomeric pair states. The word "isomer" was associated with excited nuclear states only in 1934 [5], in analogy to the chemical counterpart, interpreted as physical re-arrangement of the basic constituents: atoms for molecules and nucleons in the case of nuclei.

The definition of what could be accurately referred to as an isomer, particularly concerning the lifetime of the excited nuclear state, has changed over time [4, 6]. To facilitate the temporal and spatial isolation of nuclear excited states from the prompt radiation arising from production processes, experimental half-lives ranging from picoseconds to nanoseconds were required, depending on the particular decay product under observation. Lifetime measurements were initially limited by electronics and the definition of isomer followed the technological evolution hand in hand. In 1997, the first version of the nuclear database NUBASE considered isomers as those excited states with half-lives longer than 1 ms [7]. Later on, in 2003 [8], the same database adopted the threshold of 100 ns, although there is no fundamental reason not to extend the definition also to shorter-living excited states. The most recent evaluation of such database (2020) lists 3340 nuclides and 1938 isomeric states having a half-life $T_{1/2} \geq 100$ ns [9]. If a lower limit is considered for the definition of isomer ($T_{1/2} \geq 10$ ns), as proposed by the Atlas of Nuclear Isomers [10], the number of

S. Gargiulo, *Electromagnetic Processes of Nuclear Excitation*, Springer Theses,
https://doi.org/10.1007/978-3-031-56262-4_2

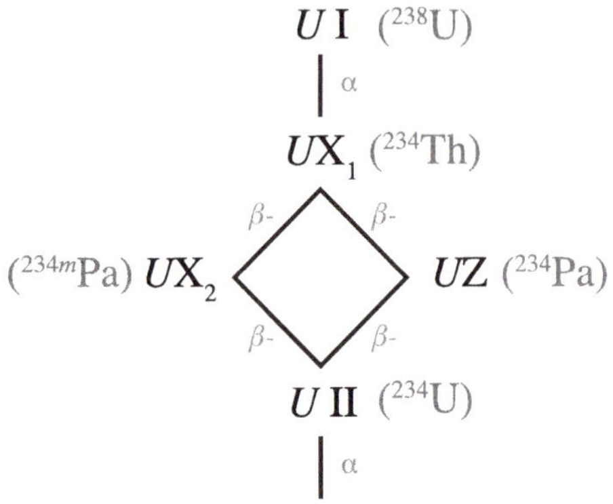

Fig. 2.1 First discovered isomer pair. Pattern of disintegration at the beginning of the uranium radioactive series with the isomeric pair UX_2 and UZ and modern identifications. The figure is redrawn from the original, published in Ref. [1]

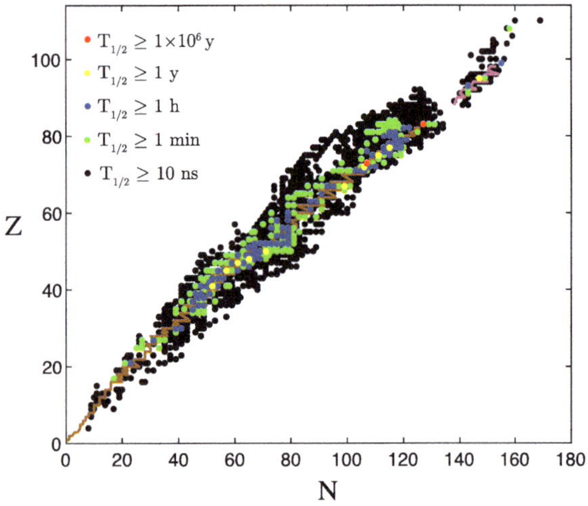

Fig. 2.2 Isomeric states as a function of the neutron and proton numbers along the whole nuclear chart. Half-lives are split into five categories, distinguished by different colors. The line of beta-stability is composed of two parts: stable nuclei in brown and β-stable nuclei with half-lives higher than 100 years in magenta

identified excited states becomes 2623 as of November 2022. The occurrence of isomers with $T_{1/2} \geq 10$ ns across the whole nuclear chart is presented in Fig. 2.2, using data extracted from the LiveChart of Nuclides [11]. The total number of retrieved isomers is 2367, which is slightly lower than the number reported by the Atlas [10], which uses multiple databases.

Figure 2.3 reveals that the quasi-totality (2091) of the nuclear excited states have a lifetime smaller than 1 min, and the number of levels rapidly declines as the timescale increases, which is expected. Of the total number of states, only 99 isomers have a lifetime longer than one hour, and merely two isomers have a lifetime that exceeds one million years. Notably, isomers with longer lifetimes are predominantly situated in the vicinity of the line of β-stability. Although isomers can be found throughout the nuclear chart, from the lightest isomer represented by ^{12}Be to the heaviest isomer given by ^{279}Ds, it is evident that they tend to accumulate in proximity to the proton and neutron magic numbers, as depicted in Fig. 2.4. In addition, also the energy landscape offers a broad spectrum of possibilities. Isomers are found to exist across

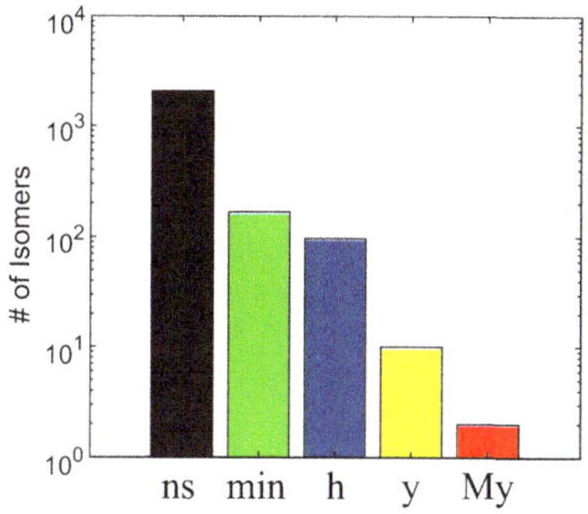

Fig. 2.3 Histogram of the isomer half-lives. Color code follows the one shown in Fig. 2.2. The label "ns" includes all the isomer in the interval 10 ns–1 min, the label "min" those in the interval 1 min–1 h and so on

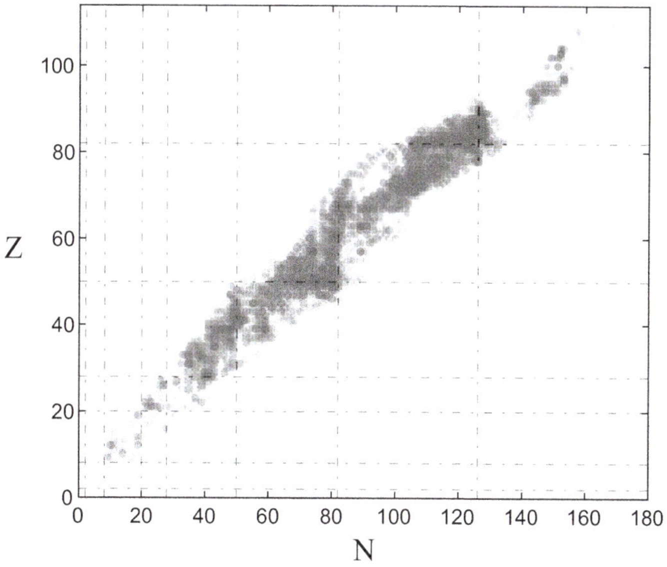

Fig. 2.4 Scatter plot of the distribution of isomers along the nuclear chart. The dashed vertical and horizontal lines represent the neutron and proton magic numbers, respectively. The markers are rendered with a transparency of 75%; hence, darker markers indicate a higher density of plotted points

a range of energies, from the low-lying 8 eV state of ^{229m}Th and the 76 eV state of ^{235m}U to the 13.67 MeV energy level of ^{208m}Pb. Figure 2.5 shows a 3D scatter plot that provides an overview of the energy and half-life distribution of isomers as a function of proton and neutron numbers. Moreover, the lateral panels incorporate information on the nuclear total angular momentum I.

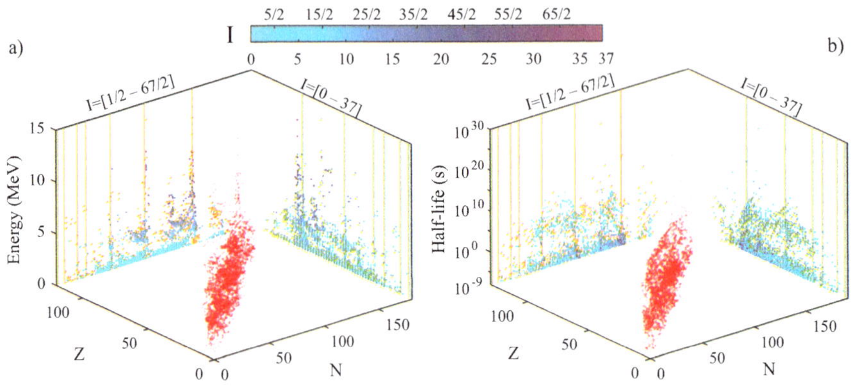

Fig. 2.5 Excitation energies and half-lives of isomers on the nuclear chart. Occurrence of isomers on the nuclear chart with respect to the energy of excitation (**a**) and half-life (**b**). Red markers are rendered with 75% transparency, while projections on the side panels are fully opaque. The vertical orange and green lines correspond to neutron and proton magic numbers, respectively. The lateral panels are color-coded as follows: (i) all isomers are represented by orange (projection on the neutron axis) or green markers (projection on the proton axis); (ii) lateral panels involving the neutron axis display the total angular momentum I of half-integer spin nuclei following the dedicated colormap; (iii) lateral panels involving the proton axis display the total angular momentum I of integer spin nuclei. The default color (orange or green) has been retained for isomers for which I was not defined. In cases of multiple I assignments, the minimum value has been chosen

The tendency of high-energy states to congregate in proximity to the magic numbers is a result of the shell structure effects. As it was initially described in 1949 [12], a strong correlation exists between the appearance of isomers in odd-A nuclei and the occupancy of intruder orbitals [13, 14]. This tendency gives rise to the formation of the so-called "islands of isomerism". As the magic numbers are approached, the energy levels associated with two consecutive harmonic oscillator shells, with principal quantum numbers N and N−1, begin to compete in energy, as illustrated in Fig. 2.6. This competition leads to the formation of nuclear shells composed of levels having all the same parity (principal quantum number N−1), with the addition of the level of highest spin of the immediately higher harmonic oscillator shell (principal quantum number N), which has the opposite parity. This unique nature of high spin difference and opposite parity makes intruder orbitals play a significant role in the emergence of isomerism. Moreover, the energy levels at which these isomers lie tend to increase close to the magic numbers, as illustrated in Fig. 2.5a. The three-dimensional plots presented in Fig. 2.5 also offer a comprehensive view of the distribution of the total angular momentum I, commonly referred to as nuclear spin, across the entire nuclear chart. Extremely high-I isomers are observed again in correspondence with the magic number and are characterized by energies on the order of several MeV. For half-integer spin nuclei, the highest value reported is ($67/2^-$) for the excited state at 10.28 MeV of ^{151}Er with a half-life $T_{1/2} = 0.42\,\mu$s. Overall ^{212}Rn has the highest spin state isomer (37^-) at 12.21 MeV and $T_{1/2} = 17$ ns. As evidenced

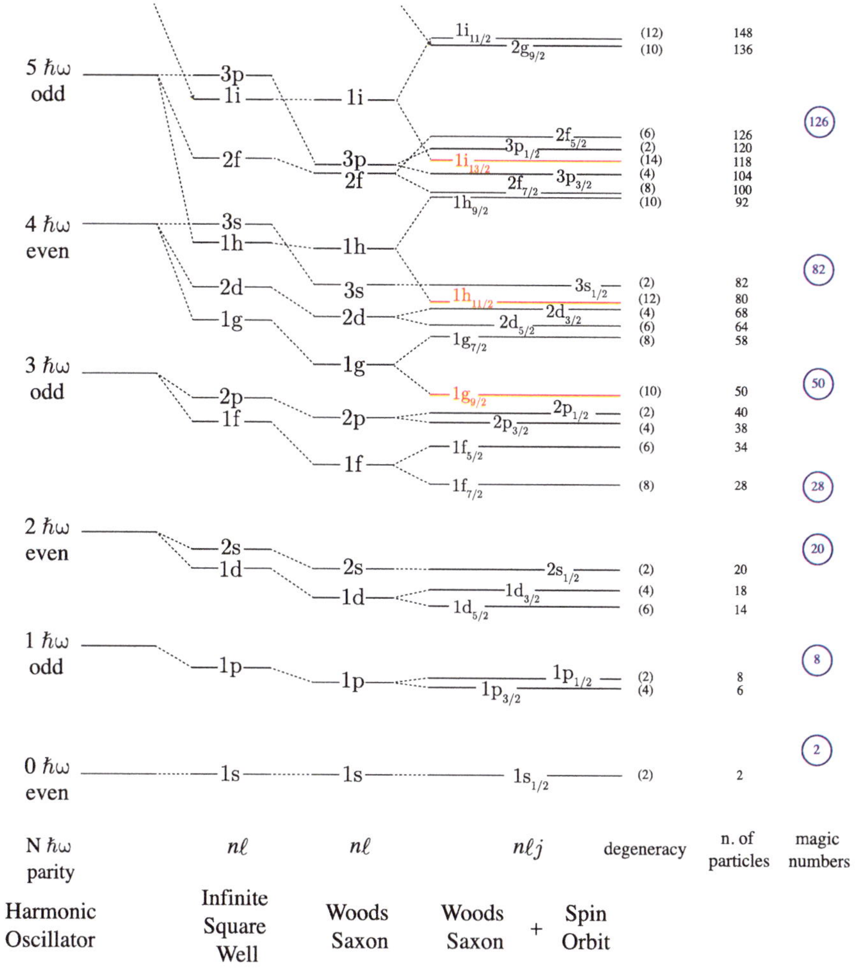

Fig. 2.6 Nuclear shell model and intruder orbitals. Schematic energy levels of the nuclear shell model in case of harmonic oscillator potential, infinite square-well, Woods-Saxon potential [15] and Woods-Saxon plus spin-orbit coupling. Levels of degeneracy and magic numbers are displayed on the right. Intruder orbitals, highlighted in orange, exhibit opposite parity in comparison to the other members of the nuclear shell, being pushed below the magic gaps by the spin-orbit coupling. The figure is redrawn and adapted from Ref. [16]

from Fig. 2.5b, these high-I nuclei generally have a lifetime up to a fraction of ms, as for the case of ^{179}Re with a $T_{1/2} = 0.46$ ms and spin-parity $I^{\pi} = (47/2, 49/2^{-})$ at 5.41 MeV. In spite of their high spin values, the limited half-life of these high-I nuclear levels can be attributed to their high energies. As these levels are in the MeV region, it is highly probable for the isomer to have several underlying levels with similar spins that are connected by low-order multipolar γ-transitions. On the other hand, the lowest possible value for the angular momentum is 0. Thirty-five special cases of $I = 0$ spin isomers can be identified from the LiveChart database, which have been reported in Appendix A.1. Among these, 18 cases lead to the lowest possible multipolar transition (E0) towards a 0^{+} ground state in even-even nuclei (reported in Appendix A.2). Moving to the other extreme of the half-life distribution depicted in Fig. 2.5b, there is ^{180m}Ta with $T_{1/2} \geq 4.5 \times 10^{16}$, which is Earth's only naturally occurring isomer and nature's rarest primordial nuclide [17]. Compared to the high-I states described earlier, the 77.2 keV isomer of ^{180m}Ta has a distinct behavior. With a spin of 9^{-} and relatively low energy, this isomer only has two lower-lying states: the 1^{+} ground state and the 2^{+} excited state. Thus only a transition involving a significant change in the nuclear spin could enable its decay. As will become clear later, these transitions are highly forbidden.

In addition to all the possible decay modes available to nuclei, such as α-decay, β-decay, proton decay, and spontaneous fission, isomers can decay electromagnetically either by γ-emission, internal conversion (IC), and internal $e^{+}e^{-}$ pair production [18, 19]. These processes (γ, IC and $e^{+}e^{-}$) are generally referred to as isomeric transitions (IT) and are unavailable decay modes for the ground state. The appearance of isomers in nuclei is itself a manifestation of a mechanism that inhibits or hinders the transition of the nucleus to lower-lying states. This hindrance is often quantified using the Weisskopf hindrance factor, defined as [20]:

$$F_{\mathrm{W}} = \frac{T^{\gamma}_{1/2}(\text{experimental})}{T^{\mathrm{W}}_{1/2}(\text{theoretical})}, \tag{2.1}$$

where $T^{\gamma}_{1/2}$ is the partial γ-decay half-life and $T^{\mathrm{W}}_{1/2}$ is the theoretical partial γ-decay half-life obtained using Weisskopf single-particle estimates [21, 22]. The probability of absorbing or emitting a photon of type λ (electric or magnetic) with energy $E_{\gamma} = \hbar ck = \hbar\omega$ and angular momentum quantum numbers L and m, as a result of a transition between two nuclear states represented by their wavefunctions Ψ_{i} (initial) and Ψ_{f} (final), can be expressed as [23, 24]:

$$W_{\mathrm{fi}}(\lambda, Lkm) = \frac{8\pi(L+1)}{\hbar\, L\, ((2L+1)!!)^2} \left(\frac{E_{\gamma}}{\hbar c}\right)^{2L+1} \left|\left\langle \Psi_{\mathrm{f}} \left| \hat{\mathfrak{M}}(\lambda, Lkm) \right| \Psi_{\mathrm{i}} \right\rangle\right|^2, \tag{2.2}$$

where $\hat{\mathfrak{M}}(\lambda, Lkm)$ is the multipole transition operator. Generally, measurements do not distinguish between different orientations m of the angular momentum L, thus one has to average on the initial orientations m_{i} of I_i and sum over all final states m_{f}

$$W_{\mathrm{fi}}(\lambda, L) = \frac{1}{2I_{\mathrm{i}}+1} \sum_{m_{\mathrm{i}}, m_{\mathrm{f}}} W_{\mathrm{fi}}(\lambda, kLm) \,. \tag{2.3}$$

Using the Wigner-Eckart theorem, it occurs the definition of reduced matrix elements [25]:

$$\left\langle I_{\mathrm{f}} m_{\mathrm{f}}, \Psi_{\mathrm{f}} \left| \hat{\mathfrak{M}}(\lambda, Lkm) \right| I_{\mathrm{i}} m_{\mathrm{i}}, \Psi_{\mathrm{i}} \right\rangle = (-1)^{I_{\mathrm{f}}-m_{\mathrm{f}}} \times \\ \times \begin{pmatrix} I_{\mathrm{f}} & L & I_{\mathrm{i}} \\ -m_{\mathrm{f}} & m & m_{\mathrm{i}} \end{pmatrix} \left\langle I_{\mathrm{f}}, \Psi_{\mathrm{f}} \left\| \hat{\mathfrak{M}}(\lambda, L) \right\| I_{\mathrm{i}}, \Psi_{\mathrm{i}} \right\rangle . \tag{2.4}$$

The sum over m_{i} and m_{f} can be performed taking advantage of the invariance under even permutation and the orthogonality conditions of the 3-j Wigner symbols [25]:

$$\sum_{m_{\mathrm{i}}, m_{\mathrm{f}}} \begin{pmatrix} I_{\mathrm{i}} & I_{\mathrm{f}} & L \\ m_{\mathrm{i}} & -m_{\mathrm{f}} & m \end{pmatrix}^2 = \frac{1}{2L+1}(2L+1) = 1 \,, \tag{2.5}$$

since there are only $(2L+1)$ value pairs $(m_{\mathrm{i}}, m_{\mathrm{f}})$ that satisfy the condition $m_{\mathrm{i}} - m_{\mathrm{f}} + m = 0$ for which the Wigner 3-j symbol is not zero. Therefore, the probability (transition rate) of radiative nuclear transition between initial and final states can be expressed as:

$$W_{\mathrm{fi}}(\lambda, L) = \frac{8\pi(L+1)}{\hbar L((2L+1)!!)^2} \left(\frac{E_\gamma}{\hbar c}\right)^{2L+1} B(\lambda L, I_{\mathrm{i}} \to I_{\mathrm{f}}) \,. \tag{2.6}$$

$B(\lambda L, I_{\mathrm{i}} \to I_{\mathrm{f}})$ is called "reduced transition probability" and can be defined in terms of the reduced matrix element of the electromagnetic multipole operator for an electric or magnetic transition:

$$B(\lambda L, I_{\mathrm{i}} \to I_{\mathrm{f}}) = \frac{1}{2I_{\mathrm{i}}+1} \left| \left\langle I_{\mathrm{f}}, \Psi_{\mathrm{f}} \left\| \hat{\mathfrak{M}}(\lambda, L) \right\| I_{\mathrm{i}}, \Psi_{\mathrm{i}} \right\rangle \right|^2 \,. \tag{2.7}$$

The Wigner 3-j symbol also defines the selection rules for the nuclear transition as follows:

$$|I_{\mathrm{i}} - I_{\mathrm{f}}| \leq L \leq |I_{\mathrm{i}} + I_{\mathrm{f}}| \qquad \text{with} \quad L > 0, \tag{2.8}$$

$$m_{\mathrm{f}} - m_{\mathrm{i}} = m \,. \tag{2.9}$$

Since the computation of the matrix element is ultimately related to the calculation of an integral of the form $\int \Psi_{\mathrm{f}}^* \, \hat{\mathfrak{M}} \, \Psi_{\mathrm{i}} \, d^3r$, an overall even-parity is required for the transition not to vanish:

$$\pi_{\mathrm{i}} \, \pi_{\hat{\mathfrak{M}}_{\lambda L}} \, \pi_{\mathrm{f}} = 1 \,, \tag{2.10}$$

with $\pi_{\hat{\mathfrak{M}}_{\lambda L}}$ being the parity of the multipole electric or magnetic radiation

$$\pi_{\hat{\mathfrak{M}}_{EL}} = (-1)^L \quad \text{and} \quad \pi_{\hat{\mathfrak{M}}_{ML}} = (-1)^{L+1} . \tag{2.11}$$

These selection rules imply that transitions of E and M types cannot occur with the same multipolarity L between the same pair of nuclear states. If a parity change is involved between the initial and final states ($\Delta\pi$ = yes), the radiation field must have odd parity, while when no change occurs ($\Delta\pi$ = no), the field must have even parity [26]. Hence, depending on the angular momenta of the states, one can have either E1, M2, E3 or M1, E2, M3, with their relative higher orders. Monopole E0 transitions ($L = 0$) are not allowed to decay via single photon emission since the intrinsic spin of a photon is 1, and the total angular momentum transferred to it by the decay is $L\hbar$. Pure E0 decays, which correspond to transitions between I = 0 states possessing the same parity—reported in Table A.2 in the case of isomers—usually proceed through the internal conversion process, in which an orbital electron is ejected from the atom into the continuum. When the excitation energies are greater than or equal to twice the mass of the electron ($\geq 2m_e = 1022$ keV), internal pair formation (IPF) becomes significant for E0 transitions. Alternative E0 decay modes involve two-quantum processes, such as pair γ-emission. However, the probabilities of these processes are generally very low, with branching ratios on the order of $\sim 10^{-4}$ [27, 28].

Calculation of the matrix elements requires knowledge of the nuclear wavefunctions, which is generally the most challenging task. Weisskopf introduced rough estimates in the single-particle picture, where a nuclear excitation is due to a nucleon moving from one orbit to another, leaving the rest unaltered. In the limit of long wavelengths, i.e. the wavelength of the radiation is large with respect to the nuclear radius ($E_\gamma \ll 197\ A^{-1/3}$ MeV [23]), it is possible to obtain the following expression for an electric multipole [23]:

$$\begin{aligned}\left\langle I_f, \Psi_f \left\| \hat{\mathfrak{M}}(E, L) \right\| I_i, \Psi_i \right\rangle &= \left\langle I_f, \Psi_f \left\| e r^L Y_L(\theta, \phi) \right\| I_i, \Psi_i \right\rangle \\ &= e \frac{1 + (-1)^{l_i + l_f + L}}{2} \left\langle f \left| r^L \right| i \right\rangle \sqrt{\frac{(2I+1)(2I_f+1)(2I_i+1)}{4\pi}} \\ &\times (-1)^{I_f - 1/2} \begin{pmatrix} I_f & I & I_i \\ -1/2 & 0 & 1/2 \end{pmatrix} ,\end{aligned} \tag{2.12}$$

where the radial integral $\left\langle f \left| r^L \right| i \right\rangle$ is defined as

$$\left\langle f \left| r^L \right| i \right\rangle = \int_0^\infty R_f(r) r^L R_i(r) r^2 \mathrm{d}r , \tag{2.13}$$

with $R_i(r)$ and $R_f(r)$ being the radial part of the nuclear wavefunctions. A similar expression can be written for a magnetic multipole. Weisskopf obtained an estimate of the radial integral by replacing $R_i \sim R_f$ with a constant up to the nuclear radius ($r \leq R_0$) and zero outside:

$$\langle f \left| r^L \right| i \rangle = \int_0^{\infty} R_{\mathrm{f}}(r) r^L R_{\mathrm{i}}(r) r^2 \mathrm{d}r \sim \frac{\int_0^{R_0} r^L r^2 \mathrm{d}r}{\int_0^{R_0} r^2 \mathrm{d}r} = \left(\frac{3}{L+3}\right) R_0^L . \tag{2.14}$$

In this expression, the integral in the denominator has been included for normalization [26]. Furthermore, considering $I_{\mathrm{i}} = L + 1/2$, $I_{\mathrm{f}} = 1/2$, $l_{\mathrm{i}} = L$ and $l_f = 0$ [22, 23], the following expressions were obtained for the reduced transition probabilities:

$$\begin{aligned} B_{\mathrm{W}}(EL) &\simeq \frac{(1.2)^{2L}}{4\pi} \left(\frac{3}{L+3}\right)^2 A^{2L/3} \quad e^2 \cdot fm^{2L} , \\ B_{\mathrm{W}}(ML) &\simeq \frac{10}{\pi} (1.2)^{2L-2} \left(\frac{3}{L+3}\right)^2 A^{(2L-2)/3} \quad \mu_N^2 \cdot fm^{2L-2} , \end{aligned} \tag{2.15}$$

where the formula $R_0 = 1.2 A^{1/3}$ fm has been used for the nuclear radius [24, 29]. The reduced transition probability is given in units of $e^2 \cdot \mathrm{fm}^{2L}$ for electric multipoles and in units of $\mu_N^2 \cdot \mathrm{fm}^{2L-2}$ for magnetic multipoles. Finally, Eq. 2.6 can be rewritten for the two types of monopoles as follows [24]:

$$\begin{aligned} W_{\mathrm{fi}}(E, L) &= \alpha \hbar c \, \frac{8\pi(L+1)}{\hbar L((2L+1)!!)^2} \left(\frac{E_\gamma}{\hbar c}\right)^{2L+1} \quad B(EL, I_{\mathrm{i}} \rightarrow I_{\mathrm{f}}) , \\ W_{\mathrm{fi}}(M, L) &= \alpha \hbar c \left(\frac{\hbar}{2M_{\mathrm{p}}c}\right)^2 \frac{8\pi(L+1)}{\hbar L((2L+1)!!)^2} \left(\frac{E_\gamma}{\hbar c}\right)^{2L+1} \quad B(ML, I_{\mathrm{i}} \rightarrow I_{\mathrm{f}}) , \end{aligned} \tag{2.16}$$

where the Gaussian units relations $e^2 = \alpha \hbar c$ and $\mu_{\mathrm{N}} = \frac{e\hbar}{2M_{\mathrm{p}}c}$ have been used to convert the numerical values in SI units. Table 2.1 lists the transition rates and Weisskopf's single-particle estimates of the reduced transition probability for the first four multipole orders.

Considering a medium-weight nucleus with $A = 125$ and a transition energy of $E = 1$ MeV, electric radiation is about two orders of magnitude more likely than magnetic multipole transition for a given order L, despite never competing for the

Table 2.1 Transition rates W (s^{-1}) and Weisskopf's estimates of the reduced transition probabilities $B_{\mathrm{W}}(\lambda L)$ for the electric ($e^2 \cdot fm^{2L}$) and magnetic ($\mu_N^2 \cdot fm^{2L-2}$) multipoles. Energies E are in units of MeV. For physical constants, the 2018 CODATA recommended values have been used [30]

	Electric		Magnetic	
L	W (s^{-1})	$B_{\mathrm{W}}(EL)$	W (s^{-1})	$B_{\mathrm{W}}(ML)$
1	$1.590\times10^{15} E^3 B(E1)$	$6.446 \times 10^{-2} A^{2/3}$	$1.758\times10^{13} E^3 B(M1)$	1.790
2	$1.225\times10^{9} E^5 B(E2)$	$5.940 \times 10^{-2} A^{4/3}$	$1.355\times10^{7} E^5 B(M2)$	$1.650 A^{2/3}$
3	$5.708\times10^{2} E^7 B(E3)$	$5.940 \times 10^{-2} A^{2}$	$6.312\times10^{0} E^7 B(M3)$	$1.650 A^{4/3}$
4	$1.697\times10^{-4} E^9 B(E4)$	$6.285 \times 10^{-2} A^{8/3}$	$1.876\times10^{-6} E^9 B(M4)$	$1.746 A^{2}$

Table 2.2 Parity-favored and parity-unfavored transitions. The table shows the lowest possible multipoles for electric and magnetic radiations from the initial I_i, π_i to the final nuclear state I_f, π_f. This table is adapted from Ref. [22]

$I_i \neq I_f$		
Parity-favored	Electric	Magnetic
$\pi_i\pi_f = (-1)^{I_i-I_f}$	$L = \|I_i - I_f\|$	$L = \|I_i - I_f\| + 1$ none if I_i or $I_f = 0$
Parity-unfavored		
$\pi_i\pi_f = (-1)^{I_i-I_f+1}$	$L = \|I_i - I_f\| + 1$ none if I_i or $I_f = 0$	$L = \|I_i - I_f\|$
$I_i = I_f \neq 0$		
	Electric	Magnetic
$\pi_i = \pi_f$	L = 2 none if $I_i = I_f = 1/2$	L = 1
$\pi_i = -\pi_f$	L = 1	L = 2 none if $I_i = I_f = 1/2$

same transition due to selection rules. The probability of emission rapidly decreases with higher multipole orders. Within the first ten multipoles, two consecutive electric transitions differ by four to five orders of magnitude, as do two consecutive magnetic transitions. Therefore, a given transition is generally dominated by the two lowest multipole orders L. The two lowest multipole orders allowed by angular momentum and parity selection rules are reported in Table 2.2.

In parity-favored transitions, the electric radiation of order L competes with the magnetic radiation of the next higher order ($L' = L + 1$), while the inverse is true in the case of parity-unfavored transitions. The estimates presented in Table 2.1 lead to the following relations:

$$\frac{W_W(EL')}{W_W(EL)} \simeq \frac{W_W(ML')}{W_W(ML)} \simeq 10^{-5}\,, \tag{2.17}$$

$$\begin{aligned} &\text{parity-favored} && \frac{W_W(ML')}{W_W(EL)} = \frac{W_W(ML')}{W_W(ML)}\frac{W_W(ML)}{W_W(EL)} \simeq 10^{-5}\,10^{-2} = 10^{-7}\,,\\ &\text{parity-unfavored} && \frac{W_W(EL')}{W_W(ML)} = \frac{W_W(EL')}{W_W(EL)}\frac{W_W(EL)}{W_W(ML)} \simeq 10^{-5}\,10^{2} = 10^{-3}\,. \end{aligned} \tag{2.18}$$

In the previous equations, $W_W(\lambda L)$ represents the transition rate obtained using Weisskopf's single-particle estimates. These estimates predict purely electric radiation in parity-favored transitions and purely magnetic radiation in parity-unfavored transitions, although with very different relative strengths. While magnetic radiation is much smaller than electric radiation in parity-favored transitions, the electric multipole competes more effectively with the magnetic one in parity-unfavored transitions. Actual transition rates can vary by several orders of magnitude due to the specific nuclear properties and nature of the excitation mode. While parity-favored transitions are experimentally found to be usually purely electric, it is possible to have $W(EL') \geq W(ML)$ in parity-unfavored transitions, despite Weisskopf's estimates predicting primarily magnetic radiation. Therefore, Weisskopf's estimates are intended to be an orientation tool to unveil the physics behind the transition rather than a proper theoretical estimate of experimental values. A much smaller γ-decay rate than the one predicted by the estimates could indicate a poor overlap between the initial Ψ_i and final Ψ_f nuclear wavefunctions, while a larger rate might indicate that more than one single nucleon is responsible for the transition, acting collectively and leading to vibrational or rotational excitation modes [24]. For this reason, Weisskopf's estimates are used as a unit (referred to as W.u.) to express the experimentally measured reduced transition probabilities.

2.1 Classification and Nature of Isomers

Isomers can be classified into three main categories depending on the specific mechanism hindering their decay, which can occur in any combination: spin isomers in spherical nuclei, K-isomers in axially-deformed nuclei, and shape isomers in nuclei exhibiting shape-coexistence [4, 6]. Figure 2.7 provides a schematic visualization of the nuclear potential energy as a function of different nuclear variables such as spin, spin projection, and deformation for the three categories of isomers.

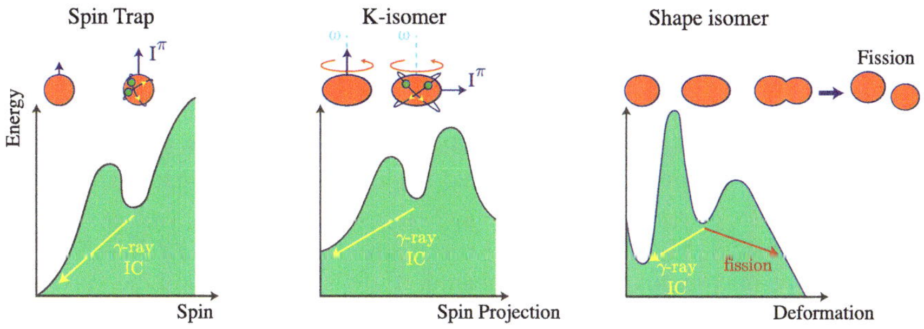

Fig. 2.7 Classification of nuclear isomers. Nuclear potential energy as a function of spin variation, spin projections, and shape deformation. The blue arrows denote the nuclear spin (I^π), while ω indicates the rotational axis. The figure has been redrawn and adapted from Refs. [31, 32]

2.1.1 Spin Traps

The nature of spin traps reflects the findings summarized in Table 2.1, which show a strong suppression of transitions with a significant change in nuclear spin (ΔI) and low energy. These circumstances generally favor the competition of internal conversion for the isomer decay. Neglecting higher-order mechanisms, the total transition probability becomes:

$$W_{\mathrm{T}} = W_{\gamma} + W_{\mathrm{IC}} = W_{\gamma}(1 + \alpha_{\mathrm{IC}}), \tag{2.19}$$

where α_{IC} is the internal conversion coefficient (ICC). The actual half-life and lifetime of these excited nuclear levels are defined in terms of the total transition probability W_{T} as follows [24]:

$$\text{Half-life } T_{1/2} = \frac{\ln 2}{W_{\mathrm{T}}}; \quad \text{Lifetime } \overline{T} = \frac{\int_0^{\infty} t e^{-W_{\mathrm{T}}t} \mathrm{d}t}{\int_0^{\infty} e^{-W_{\mathrm{T}}t} \mathrm{d}t} = \frac{1}{W_{\mathrm{T}}} = \frac{T_{1/2}}{\ln 2}. \tag{2.20}$$

The explanation of isomerism in terms of forbidden γ-transition due to large spin variation was first proposed by Weizsäcker in 1936 [33]. In this vision, isomerism emerges whenever a large ΔI exists between a state and all other levels lying below it. It is generally possible to talk about spin traps when the transition involves a $\Delta I = L \geq 3$, where Eq. 2.16 predict a radiative half-life $T_{1/2}^{\gamma} \geq 1$ min when $E \sim 80$ keV.

In this context, Racah [34] and Flowers [35] introduced an additional quantum number, the seniority υ, to indicate the number of unpaired nucleons for a given nuclear state. In even-even nuclei, all nucleons are paired, resulting in a seniority of $\upsilon = 0$ and a ground state of $I = 0^+$. However, in odd-A and odd-odd nuclei, the unpaired nucleons (one for odd-A and two for odd-odd) are expected to determine the properties of the state, such as the angular momentum. Therefore, lower-lying excited states in such nuclei can be interpreted as transitions of odd particles between orbitals of the same shell. The representation of this interpretation can be seen in Fig. 2.7, where spin traps are drawn as two excited nucleons orbiting around a spherically symmetric core.

In odd-A nuclei, isomerism is not expected to occur up to Z or N = 20. This is because the maximum angular momentum of the shells is $5/2$ and the maximum $\Delta I = 2$, as seen in Fig. 2.6. When $20 \leq (Z, N) \leq 38$, the levels $f_{7/2}$, $f_{5/2}$, and $p_{3/2}$ are being filled with the highest possible multipolar transition still of the order $L = 2$. When $(Z, N) \geq 38$, the orbitals $g_{9/2}$ and $p_{1/2}$ compete for the ground state with a spin difference of $\Delta I = 4$, resulting in the emergence of isomeric states. Thus, the nuclear shell model predicts the first "island of isomerism" in the range $39 \leq (Z, N) \leq 49$ [14, 22]. After shell closure at the magic number 50, competition begins between the intruder orbital $g_{7/2}$ and the level $d_{5/2}$, which are completely filled for $(Z, N) = 64$. Since $\Delta I = 2$, no long-lived excited states are expected in this range; however, isomerism can emerge just before $(Z, N) = 64$ due to transitions to the empty higher levels ($h_{11/2}$ or $d_{3/2}$), which are relatively close in energy. The occurrence of isomers

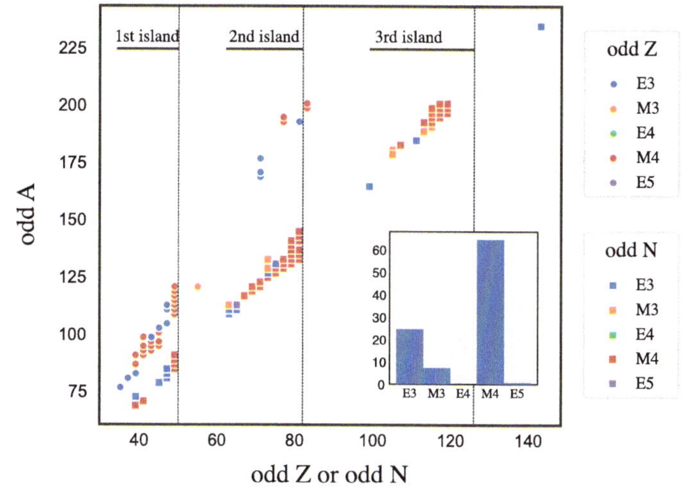

Fig. 2.8 Isomeric islands in odd-A nuclei. Distribution of isomers occurring in odd-A nuclei with $T_{1/2} \geq 60$ s and $E < 1$ MeV plotted as a function of the odd nucleon number (Z or N). The color code of each marker in the plot represents the lowest multipole order radiated by the isomeric level. When multiple transitions are possible (i.e., several decays towards lower-lying levels), the electric multipole has been preferred over the magnetic for the assignment in case of equal order L (see Eq. 2.18). For the remaining cases, the strongest relative transition was chosen. The inset of the figure shows the histogram of occurrence with respect to the multipolar decays. The vertical dashed lines in the plot represent nuclear magic numbers. The data used to generate this figure were extracted from the LiveChart of Nuclides [11]

should increase towards the end of the shell due to the energy competition of the $h_{11/2}$ (intruder), $s_{1/2}$ and $d_{3/2}$ orbitals up to shell closure at 82, giving rise to the second "island of isomerism". In the range $82 \leq (Z, N) \leq 100$, the lowest energy state is determined by the competition between the $h_{9/2}$ and $f_{7/2}$ orbitals, which are connected at most by a quadrupolar transition. Beyond $(Z, N) = 100$, competition involves the low-spin orbitals ($p_{3/2}$, $p_{1/2}$ and $f_{5/2}$) and the intruder orbital $i_{13/2}$, leading to the emergence of the third "island of isomerism". The distribution of isomers, shown in Fig. 2.8, with half-lives $T_{1/2}$ longer than 60 s and excitation energy E below 1 MeV (to be fairly below the pairing gap), as a function of the odd nucleon number (either Z or N) reveals the three distinct islands of isomerism. It is worth noting that, as expected, the lowest multipole radiation present is an $E3$.

In the first island, comprising nuclei with $39 \leq (Z, N) \leq 49$, $M4$ and $E3$ are the lowest available transitions, with a certain prevalence of $M4$ over $E3$. This is a result of the competition between the $p_{1/2}$ orbital with odd parity and the $g_{9/2}$ orbital with even parity. The decay modes are thus parity-unfavored (see Table 2.2) with parity change and spin change of four, which leads to $M4$ transitions. However, when an odd number of nucleons occupy the orbitals, other resultant spins—rather than simply $^9/_2$—are also possible due to their coupling. Consequently, the formation of a level with a resultant angular momentum of $^7/_2$ is responsible for the appearance of

$E3$ (octupole) transitions [14]. In the second island, due to the competition between $d_{3/2}$ (even), $s_{1/2}$ (even) and $h_{11/2}$ (odd) orbitals, both $M4$ and $E5$ are possible. Additionally, there are special cases where $E3$ and $M3$ transitions occur [14]. In the third island a similar situation emerges between $i_{13/2}$ (even), $f_{5/2}$ (odd), $p_{3/2}$ (odd) and $p_{1/2}$ (odd) orbitals. A spin difference between four and six and a parity change allow for multipoles up to $M6$, although none is observed. It is also interesting to note that no $E4$ transitions are immediately found: this is predicted by the shell model since the levels in competition involve a parity change. Nevertheless, some special cases, such as ^{179}W and ^{185}W, exhibit both $E4$ and a much stronger $M3$ transition, which cannot be easily assigned using the same logic. Among the three islands, there is a strong predominance of the $M4$ (65 cases) and $E3$ (25 cases) transitions. Only one occurrence of a pure $E5$ transition is observed for ^{113}Cd, although the primary decay is through β-emission. It is worth mentioning that the $E5$ was also considered to be the highest order of multipolar decay ever observed in isomers until very recently (March 2023), when an E6 (hexacontatetrapole) 3.04 MeV γ-decay was confirmed in $^{53\mathrm{m}}$Fe [36, 37].

In even-even nuclei ($\upsilon = 0$), the situation differs from the odd-A case. Here, the excitation of a nucleon requires the breakup of a pair resulting in the creation of a hole below the Fermi level and an unpaired particle, forming what is called a two quasi-particle (2-qp) configuration. As a result, the excitation spectrum of even-even nuclei is expected to lie entirely above the pairing gap, which is of the order of $2\Delta \sim 12A^{-1/2}$ MeV, except for those states that arise from collective motions [32]. This pairing gap is typically too high to permit the occurrence of long-lived excited states.

Compared to even-even nuclei, odd-odd nuclei already have a 2-qp configuration in their ground state without the need to break a pair. This means that the excitation spectrum of odd-odd nuclei consists of two parts: one below the pairing gap due to the excitation of the two quasi-particles and one above it due to the breaking of an additional pair and the consequent formation of a 4-qp configuration. An interesting observation can be made from Fig. 2.3: the only two isomers with a half-life $T_{1/2} \geq$ 1 My are double-odd spin trap nuclei, namely $^{180\mathrm{m}}$Ta and $^{210\mathrm{m}}$Bi, both lying very close to the line of β-stability, as illustrated in Fig. 2.2.

2.1.2 K-Isomers

The second category of isomers depicted in Fig. 2.7, highlights the importance of conserving the direction of the angular momentum with respect to a symmetry axis, rather than its magnitude, during a nuclear transition. This makes already evidently clear that this hindrance mechanism belongs to nuclei that deviate from a spherical shape. In such cases, the solutions of the nuclear wave equations cannot be identified anymore with their ℓ orbital angular momentum, as was done for those derived from the spherical models shown in Fig. 2.6, since ℓ is no longer a "good" quantum number. In a deformed potential, the degeneracy of the single-particle orbit with total angular

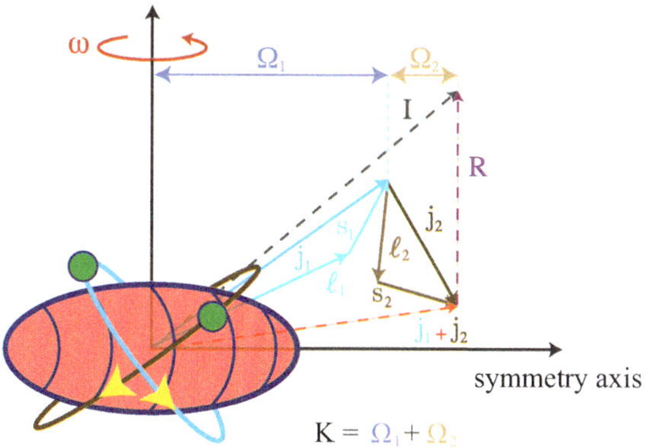

Fig. 2.9 Definition of the K quantum number as the projection of I along the symmetry axis of the deformed nucleus. In the case of a prolate spheroid, which is the most common shape of deformed nuclei, the long axis is the symmetry (deformation) axis, while the short axis is the one around which the collective motion takes place with angular speed ω. In this drawing, the total angular momentum j originates from two unpaired nucleons (2-qp excitation). The nuclear spin is given by $\mathbf{I} = \mathbf{j} + \mathbf{R}$, where $\mathbf{R}$ is the rotational angular momentum. The figure has been redrawn from Ref. [6]

momentum j is not given by $(2j + 1)$, but rather determined by its projection along the symmetry axis (denoted by Ω). Due to reflection symmetry, the degeneracy is reduced to two with $\pm\,\Omega$ sharing the same energy, while distinct projections result in different energies [26]. These considerations led to the introduction of the K quantum number in 1955 as the projection of the total angular momentum I along the symmetry axis of the deformed nucleus [38], as depicted in Fig. 2.9.

Considering its definition, it would be legitimate to expect transitions to obey the following selection rule, in addition to those in Eq. 2.11:

$$|K_\mathrm{i} - K_\mathrm{f}| \leq \Delta I \leq L\,, \tag{2.21}$$

given that a radiation of multipolarity L cannot change the angular momentum projection by more than $|L|$. In Ref. [38], the nuclear motion is divided into intrinsic (which leads to a rearrangement of the internal structure of the nucleus) and rotational (where the internal structure is preserved but rotating at different angular velocities) modes, allowing the wavefunction to be expressed as the product of two independent parts. However, deviations from this model and especially mixing between states do not make Eq. 2.21 a strict selection rule, as those in Eq. 2.11 are. Violation of Eq. 2.21 can act as a retardant for the decay, suggesting that K is only approximately conserved during the process. Interestingly, K-forbidden transitions—which violate this selection rule—are more likely to occur as long as they involve small multipole radiations rather than K-allowed transitions that require a high spin ΔI variation. In reason of these considerations, the difference between the multipolarity order L and ΔK is defined as the degree of forbiddenness:

$$v = \Delta K - L\,. \tag{2.22}$$

In continuation with the definition of the Weisskopf hindrance factor in Eq. 2.1, which aims to carry information about the multipolarity and energy of the decay, it is possible to define a reduced hindrance factor that also includes the degree of forbiddenness [20]:

$$f_v = (F_W)^{1/v}. \tag{2.23}$$

A historical example of a K-isomer is the case of ^{180m}Hf, where the first rotational band was ever observed [39–42]. The isomer in question has $I^\pi = 8^-$, excitation energy $E = 1.14$ MeV and half-life $T_{1/2} = 5.5$ h, as reported in Fig. 2.10. This state could decay towards the lower-lying 8^+ or 6^+, through the small radiation multipoles $E1$ or $M2 + E3$, respectively. However, the high value of the hindrance factor $F_W \sim 10^{16}$ cannot be explained by a simple variation of ΔI. These transitions were found to be 7-fold K-forbidden ($v = 8 - 1 = 7$). This is because the lower state of the first rotational band—where the isomer was found—has K = 8, while the lower-lying states (8^+ and 6^+) belong to the ground state rotational band and have K = 0. Thus, the E1 decay requires a variation of the angular momentum orientation from a situation where there are $8\hbar$ units of angular momentum along the symmetry axis ($I^\pi = 8^-$, K = 8) to a configuration where the angular momentum is aligned along the rotational axis ($I^\pi = 8^+$, K = 0), which is perpendicular to the axis of symmetry

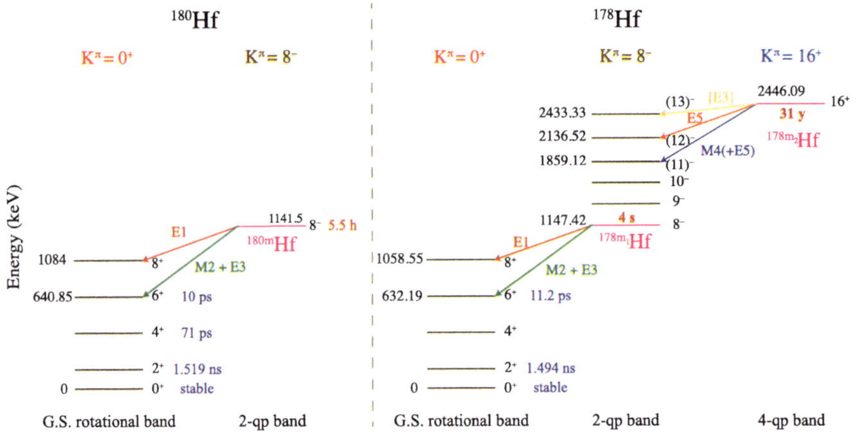

Fig. 2.10 Appearance of K-isomers in the energy level schemes of ^{180}Hf (left) and ^{178}Hf (right). Both isotopes exhibit the characteristic rotational bands of even-even nuclei. Energies, I^π, and half-lives are provided for selected levels of interest. **(left)** The ^{180m}Hf isomer is located in the 2-qp band with K = 8. The two transitions possible from this level, $E1$ and $M2 + E3$, require $\Delta K = 8 \gg L$, making them highly forbidden. **(right)** In the case of ^{178}Hf, the first isomer ^{178m_1}Hf is found in the 2-qp $K^\pi = 8^-$ band, with a half-life of 4 s. The second isomer ^{178m_2}Hf is the result of the combination of two 2-qp $K^\pi = 8^-$ excitations, leading to a 4-qp configuration of $K^\pi = 16^+$ [42]. Values in round parenthesis are uncertain, while those in square brackets are derived from theory. The data used for this figure were extracted from the ENSDF database [43]

[44], as depicted in Fig. 2.7. Löbner showed that the hindrance factor for K-isomers scales with the degree of forbiddenness and that for a large set of nuclei $f_\upsilon \sim 100$ [20]. This means that the transition rate is generally hindered by a factor of about 100 per degree of forbiddenees, i.e. $T^{\gamma}_{1/2} \sim (100)^7\, T^{\mathrm{W}}_{1/2}$ in case of $^{180\mathrm{m}}$Hf.

Another important example is the case of $^{178\mathrm{m}_2}$Hf, whose energy level scheme is shown in Fig. 2.10. This isomer is unique for its combination of high energy excitation ($E = 2.45$ MeV) and very long half-life $T_{1/2} = 31$ y [45]. The strongest decay pathway (99.9%) from this level is through a 13 keV $E3$ transition towards the lower-lying $(13)^-$ state, which is primarily accomplished through internal conversion having a coefficient $\alpha_{\mathrm{IC}} = 1.47 \times 10^7$ [46]. The appearance of this isomer can be attributed to a mixture of a spin- and K-forbidden transition, involving a spin change $\Delta I = 3$ and a variation in its orientation $\Delta K = 8$, resulting in the degree of forbiddenness $\upsilon = 5$. It is noteworthy that in the absence of such a large variation in K, the expected half-life of this state would have been only 100 ms [47], emphasizing the significance of the violation of the K-selection rule.

2.1.3 Shape Isomers

Shape isomerism arises from the existence of a secondary minimum in the potential energy landscape of the nucleus as a function of its deformations, as illustrated in Fig. 2.7. This results in the requirement of a substantial nuclear shape change for any γ-decay directed back to the ground state, potentially leading to isomerism owing to the poor overlap between the initial and final wavefunctions. According to the liquid drop model (LDM) [48], these deformations can be described by expanding the radius in spherical harmonics [16, 23, 48]:

$$R(\theta, \phi) = R_0 \left(1 + \alpha_{00} + \sum_{\lambda=1}^{\infty} \sum_{\mu=-\lambda}^{\lambda} \alpha^*_{\lambda\mu} Y_{\lambda\mu}(\theta, \phi) \right), \tag{2.24}$$

where λ is the multipolarity of the nuclear surface. The equation suggests that there are $2\lambda + 1$ variables associated with each mode of order λ. However, several conditions on symmetry, invariance, and the requirement to have $R \in \mathbb{R}$ limit the number of independent variables. Generally, any collective motions and dynamical oscillations can be described by letting $\alpha_{\lambda\mu}$ vary with time. The monopole term ($\lambda = 0$) α_{00} describes the changes in the nuclear volume, and its vibration gives rise to breathing modes. Since nuclear matter is rather stiff, the excitation levels resulting from these oscillations are significantly high in energy, typically around $\sim 80 A^{-1/3}$ MeV [24]. Therefore, assuming the incompressibility of the nuclear fluid, α_{00} can be considered fixed and expressed in terms of the other deformation parameters. For small deformations, terms with $\lambda = 1$ correspond to a translation of the entire nucleus and an oscillation of the whole system around a fixed point. The isoscalar dipole mode arises when all nucleons move together, while the isovector dipole mode arises when

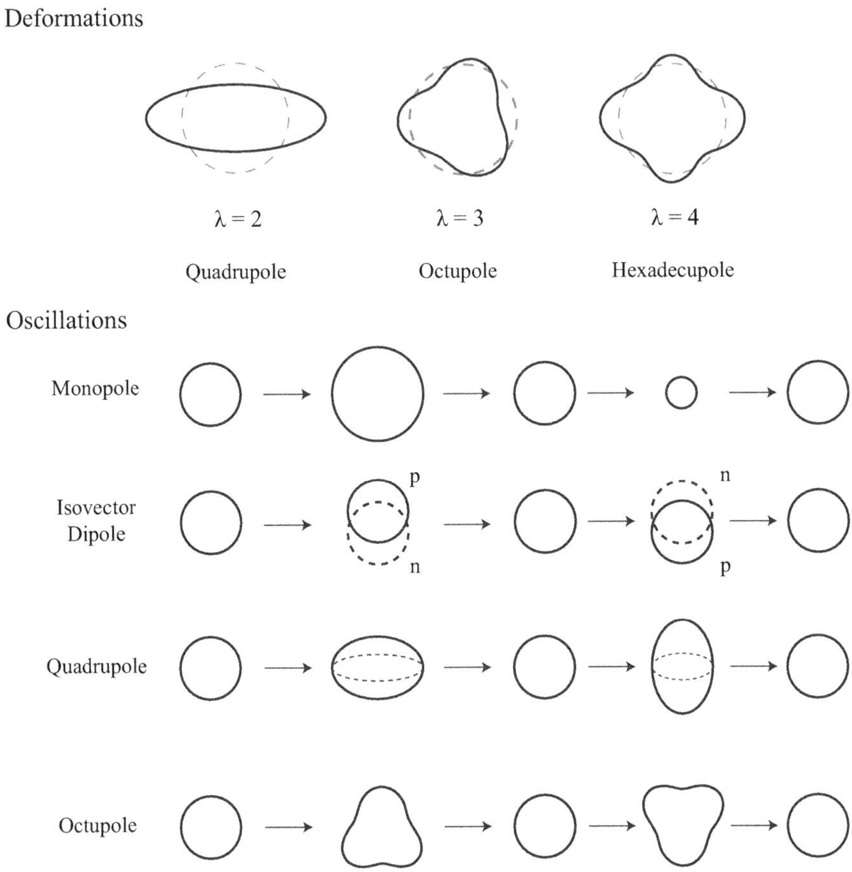

Fig. 2.11 Nuclear deformations and time evolution of vibrational modes for low-order multipoles. The figure shows the nuclear deformations corresponding to the first three non-trivial multipoles of the surface expansion and the time evolution of the corresponding vibrational modes, including $\lambda = 0, 1$. The figure has been redrawn and adapted from Ref. [24]

neutrons and protons move out of phase. The latter mode is responsible for the giant dipole resonances (GDR) [49]. However, since these first two motions are considered trivial in the analysis of deformations, their terms are generally omitted from the sum of Eq. 2.24. The first term considered in the expansion is the quadrupolar order $\lambda = 2$. Nuclear shapes and oscillations for some of the lowest multipolar orders of the expansion in Eq. 2.24 are shown in Fig. 2.11.

For quadrupolar deformations, the initial five $\alpha_{2\mu}$ coefficients reduce to a set of two new real independent variables a_{20} and a_{22} in the nucleus fixed coordinate system, where the coordinates are taken along the nucleus principal axes [16, 48]. In this context, the deformations are conveniently described in the (β, γ) space [50, 51], which is defined as follows:

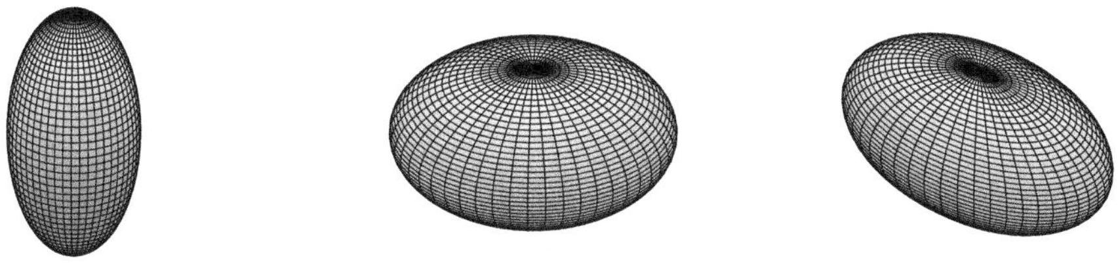

Fig. 2.12 Nuclear shapes for quadrupolar deformations.

$$a_{20} = \beta \cos \gamma$$
$$a_{22} = \frac{1}{\sqrt{2}} \beta \sin \gamma \,. \tag{2.25}$$

The values of $\gamma = [0°, 120°, 240°]$ correspond to a prolate axially symmetric deformation, while $\gamma = [60°, 180°, 300°]$ yield to an oblate axially symmetric deformation. When γ is not a multiple of 60°, the shape of the nucleus does not exhibit axial symmetry, and it is referred to as "triaxial". These deformations are illustrated in Fig. 2.12. A quadrupolar vibration involves the oscillation of the nuclear shape between these two deformations (prolate and oblate), passing through a spherical equilibrium shape $\alpha_{\lambda\mu} = 0$), as shown in Fig. 2.11. For each λ, these vibrations lead to a harmonic excitation energy spectrum if built on a spherical nucleus (no rotational degrees of freedom), while in deformed nuclei, rotational bands are built on each vibrational state [24].

The discovery of the fission process in 1939 [52, 53] was a significant milestone in nuclear physics, and the use of the liquid drop model (LDM) and its deformations played a crucial role in qualitatively describing the phenomenon. Fission occurs mainly due to the interplay between the Coulomb energy, which repels the protons and pulls them apart, and the surface energy, which tries to keep the nucleus spherical. These two factors constitute the primary contributions to the fission barrier, as depicted in Fig. 2.13. Excitation of the nucleus, as in the case of neutron capture by ^{235}U, can induce a collective motion and it may become energetically favorable for the nucleus to separate into two overlapping groups of nucleons with a distance d. The potential energy as a function of this distance presents a saddle point beyond which the fragmentation of the nucleus is inevitable. Furthermore, the inclusion of the shell structure effects leads to a double-humped fission barrier, as shown in Fig. 2.13, with a deeper minimum (I) corresponding to the ground state and the second (II) representative of shape isomerism.

Fission can be distinguished in (i) *induced-fission*, where the nucleus overcomes the barrier with sufficient energy (called activation energy), moving from a state of zero or reduced deformation to an unstable configuration; (ii) *spontaneous-fission* due to the tunneling of the barrier. The strength of spontaneous fission depends on the shape of the barrier and can occur from both potential minima, which are

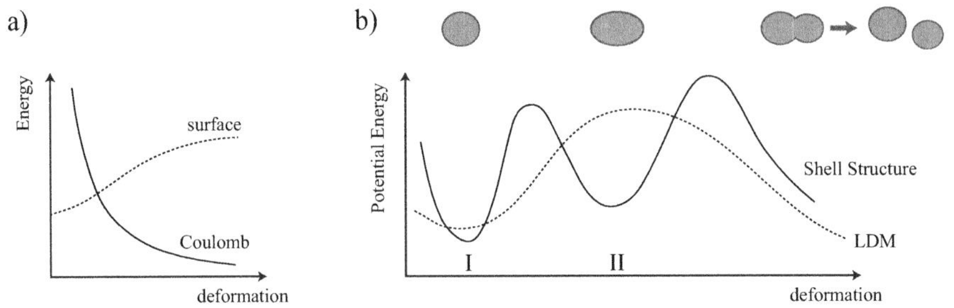

Fig. 2.13 Fission barrier and nuclear deformations. **(a)** The competition between the surface energy contribution (dashed curve) and the Coulomb term (solid curve) is shown as a function of the nuclear deformation, seen here as the distance between the two overlapping groups of nucleons. These two contributions primarily constitute the fission barrier illustrated in **(b)** with a dashed curve (LDM). The inclusion of the shell model effects results in a double-humped barrier with two minima: (I) and (II). This figure is inspired by Refs. [24, 32]

characterized by very different probabilities. Shape isomers that exhibit spontaneous fission are classified as fission isomers. The first spontaneously fissioning isomer was discovered in ^{242}Am by Polikanov et al. in 1962 [54]. ^{242}Am is known to possess two isomers: the first, denoted as ^{242m_1}Am, is located at an excitation energy of 48.60 keV and has a half-life of $T_{1/2} = 141$ y. This isomer predominantly decays via an $E4$ multipolar transition to the ground state. The second isomer—^{242m_2}Am — is found at a higher energy of 2.20 MeV, has a significantly shorter half-life of $T_{1/2} = 14$ ms and decays exclusively (100%) via spontaneous fission. It is worth noting that ^{242m_2}Am is also the longest-lived among all the known shape isomers. Currently, 48 cases of fission isomers occur in the actinide region, with $90 \leq Z \leq 98$ and $141 \leq N \leq 151$ [6]. The potential energy landscape of some nuclei can also exhibit more than two minima. For example, the nucleus ^{186}Pb has three different shapes associated with the lowest three states in the excitation spectrum: spherical, oblate, and prolate [55]. This occurrence is referred to as shape coexistence. Other examples of shape isomers are the low-lying 0^+ states in even-even nuclei, as in ^{12}Be, ^{72}Ge and ^{98}Mo.

2.2 Essence and Significance of Nuclear Isomers as Long Duration Energy Storage Solution

The previous classification might have hinted at some crucial features that make isomers attractive for both fundamental research and applications. Their excitation spectrum and the mechanisms behind their emergence provide a valuable probe for investigating nuclear properties related to shell structure and collective motions. Moreover, isomers can have a long half-life, sometimes even surpassing that of the ground state, as seen in ^{180m}Ta. In nucleosynthesis, this aspect influences the rates of nuclear reactions, leading to variations in the production and abundance of elements in the Universe [56, 57]. Other important areas of research include gamma-ray lasers [58], nuclear clocks [59], and dark matter detection [60]. Isomers also find practical applications in medical imaging. For instance, ^{99m}Tc is widely used for Single Photon Emission Computer Tomography (SPECT), while ^{34m}Cl and ^{52m}Mn are suitable for Positron Emission Tomography (PET) [61, 62].

In addition, another opportunity appears almost self-evidently when listing the properties of isomers: their ability to store energy over an extended period of time. For decades, isomers have held promise as a more compact (10^4 denser than chemical batteries) and longer-lasting energy storage solution [31, 63–65]. Long-duration energy storage (LDES) solutions have the potential to address the intermittency issues of renewable energy sources, making a cost-effective transition to decarbonized electricity possible [66]. It can be argued that the age of the Universe represents the ultimate long-duration timescale and isomers such as ^{180m}Ta—with a half-life $T_{1/2} \geq 4.5 \times 10^{16}$ y—exceed this limit [17]. The $E7$ multipolarity, which represents the lowest possible decay mode for the isomeric level at 77.2 keV, has never been observed. This suggests the absence of any form of leaking over its entire existence, a feature of utmost importance for the viability of any potential energy storage technology. Despite the small size of the single energy block stored by isomers compared to fission events ($\sim$200 MeV), the amount of energy that can be accumulated in a macroscopic sample can still be significant [31]. For example, one cubic centimeter of natural tantalum can store 82 kJ and up to 690 MJ in a pure ^{180m}Ta sample.[1] Another relevant example is represented by ^{178m_2}Hf, which exhibits a rare combination of long half-life and high excitation energy. The 2.45 MeV level can store up to 17 GJ in one cubic centimeter of pure ^{178}Hf with a half-life of 31 y.

Yet ideally, one would like not only to have the absence of leakage during the battery lifespan but also the possibility to release and re-introduce this energy "on-demand". As the main decay mode of excited nuclear states is through the isomeric transition, the electromagnetic manipulation of the isomer decay assumes particular relevance. However, isomers exist because of a hindrance mechanism that inhibits their decay; hence, it might be necessary to find alternative pathways involving upper-lying levels, to feed or deplete these energy traps.

The possibility to induce isomer depletion through direct photoexcitation, followed by the subsequent decay towards the ground state, has been demonstrated in ^{180}Ta by connecting the isomer with a set of intermediate levels through a minimum photon energy of $\sim$1 MeV [57, 67, 68], as depicted in Fig. 2.14. Belic et al. [57] found a total depopulation cross-section of 0.057 b eV for the 1.01 MeV photon, 0.27 b eV for the 1.22 MeV photon, and up to 36 b eV for photons at 2.80 MeV, which are larger than what expected from single-particle estimates [69]. While these transitions appear to bypass the repopulation of the isomer, a back-decay towards the 9^- state—that would be theoretically expected—has not yet been observed, leaving

[1] These quantities only account for the energy stored in the 77 keV isomer. ^{180}Ta has an unstable ground state with a $T_{1/2} \simeq 8$ h, thus one could consider this decay as an additional source of energy release.

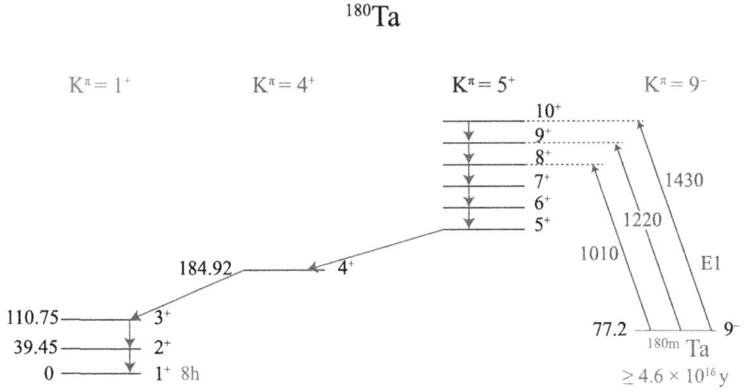

Fig. 2.14 Partial level scheme of ^{180}Ta and identification of pathways for isomer depletion. The photoexcitation pathways are indicated with red arrows, along with their photon energy in keV. It should be noted that these E1 transitions are K-forbidden ($\Delta K = 4$). The decay towards the ground state is indicated with blue arrows, while the half-lives of the isomer and the ground state are indicated in green. The figure has been redrawn and adapted from Ref. [67]

open questions [44, 69] that are further intricate by their potential astrophysical consequences [57, 70, 71]. Regardless of these considerations, the need of $E \geq 1$ MeV to initiate the isomeric energy release of 77.2 keV makes this process energetically unfavorable for practical applications in nuclear batteries.

A similar photoexcitation pattern was expected to exist for ^{178m_2}Hf, but here the situation is much more controversial, and a historical overview can be found in Ref. [72]. The first experiment, published in 1998, claimed the depletion of the ^{178m_2}Hf isomer through the absorption of photons with $E \sim 90$ keV and reported an exceptionally large total cross-section of about 10^6 beV [73]. This result stimulated considerable interest, but subsequent experiments failed to replicate the claimed depletion [74–76] and set an upper limit for the photoexcitation cross-section at ~ 2 b eV, which is more than five orders of magnitude lower than the value reported in Ref. [73]. Direct coupling of the isomeric (K $= 16^+$) and ground state (K $= 0^+$) bands, shown in Fig. 2.10, through low-order multipolar transitions ($B(\lambda L)$ ~ 1 W.u.) has been suggested to be possible, with the minimum energy requirement of 322 keV photons to activate the depletion [65, 77]. Nevertheless, up to date, these transitions have never been tested experimentally.

Besides pure photon absorption, alternative solutions for isomer depletion can be provided by atomic-assisted processes. The fact that the atomic surrounding could play a significant role in the deexcitation of the isomer was shown previously in Eq. 2.19. Most of the time, the lifetime of a state is strongly influenced by the internal conversion channel, where an orbital electron is ejected towards a free continuous state. In highly converted transitions (i.e., those with high values of α_{IC}), the half-life of the state can be tremendously increased by stripping away all atomic electrons. Therefore, at least in principle and disregarding any practical considerations, it seems

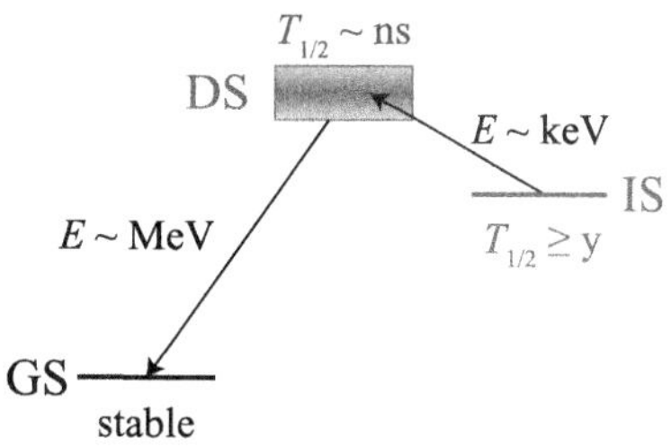

Fig. 2.15 Desirable energy level scheme for the isomer depletion. IS indicates the isomer, while DS denotes the depletion or intermediate state that facilitates the decay towards the ground state (GS)

to be always possible to *increase* the half-life. However, the opposite manipulation—i.e., the possibility to *decrease* the lifetime at will—remains an extreme challenge with ambitious applications opening up in case of success [78]. Since innermost shells contribute the most to the α_{IC}, one possibility for decreasing the lifetime could be the manipulation of the orbital electron wavefunctions via fullerene cages, as similarly done for the enhancement of the radioactive decay of ^{7}Be through electron capture [79]. However, this process is unlikely to cause an abrupt variation in the lifetime. These considerations outline several desirable ingredients for the isomer depletion scheme, illustrated in Fig. 2.15.

The first requirement is the presence of upper-lying levels that allow the decay of the isomer towards the ground state. A stable ground state is preferable if the goal is to achieve "clean" energy storage and avoid radioactive by-products or to have the ability to feed the isomer multiple times after the energy release. The lifetimes of the involved levels are also crucial: ideally, the isomer should last for years, while the depletion state should release the energy on a much faster timescale. To be energetically convenient, the energy required to activate the depletion should be significantly smaller than the energy released by the isomer. The list of mechanisms that could enable the transition between the isomeric and depletion state is extensive.

Generally, these mechanisms can be categorized as resonant photoexcitation, which involves the absorption of real photons [80], atomically assisted excitations, Coulomb excitation through inelastic scattering with charged particles, which involves the exchange of virtual photons [81], and inelastic neutron acceleration [82–84]. While Table 2.1 indicates that dipolar transitions are preferred for resonant photoexcitation, atomically assisted and Coulomb excitation processes may be more efficient for higher multipolarities. Historically, among others, two mechanisms involving atomic electrons have been extensively discussed: nuclear excitation by electron capture (NEEC) [85, 86] and nuclear excitation by electron transition (NEET) [87]. These mechanisms are intimately intermingled with the internal conversion process, as they are their time reversal (more properly, bound internal conversion is the inverse of NEET [88, 89]). Both mechanisms, NEEC and NEET, require an electronic transition to excite the nucleus. If the transition is between a continuum and a bound state, the process is called NEEC, while if it occurs between two bound orbitals, the process is referred to as NEET. NEET has been invoked as responsible for the high depletion of ^{178m_2}Hf observed in Ref. [73], but theoretical analysis reported lower cross-sections that were unable to explain the result [90–92]. While NEET is considered a well-established phenomenon, experimentally observed in ^{197}Au [93], the nature and the observation of the NEEC process is currently an active area of research, as it will be explained in detail in the next chapter. Examples of third-order processes include: (i) inverse electron bridge (IEB), where the nucleus is excited following the photon absorption by an atomic electron and the consequent transition of this electron to a level coherent with the conservation of energy [94–97]; and the (ii) nuclear excitation by two-photon electron transition (NETP), where the nucleus is excited by the virtual photon of a two photon-decay occurring in the electronic shell [98, 99]. Feynman diagrams and corresponding energy schemes depicting these mech-

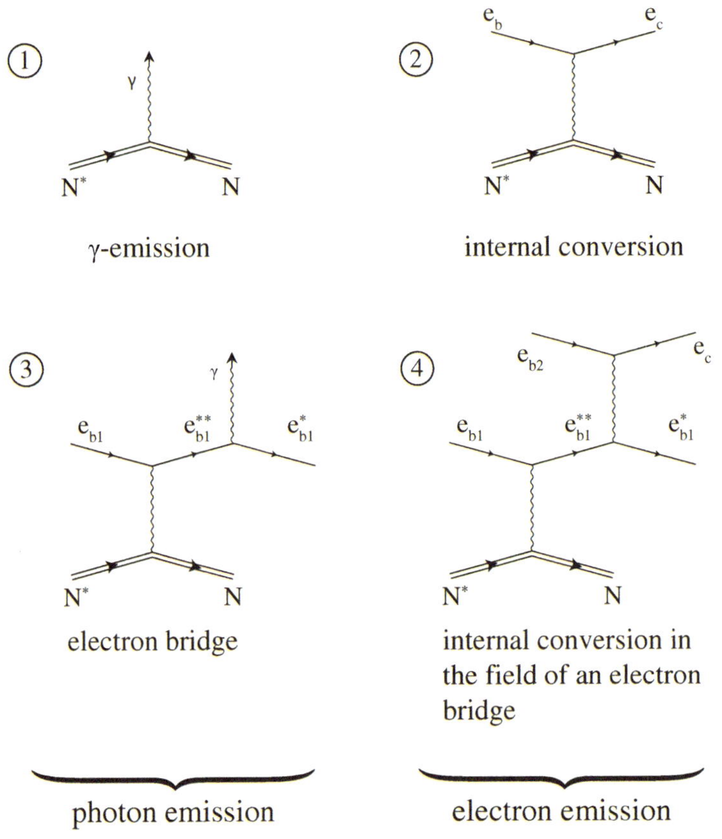

Fig. 2.16 Feynman diagrams for nuclear deexcitation processes of the first-, second-, third-, and fourth-order (in circled numbers). The nucleus is represented by the double line, while the thin single line represents an electron. Wavy lines with outgoing arrows indicate the emission of real photons, while wavy lines represent the exchange of virtual photons. The asterisk denotes excited states, while the label "b" indicates bound electronic orbitals, and "c" denotes continuum states. This figure has been redrawn and adapted from Refs. [95, 100]

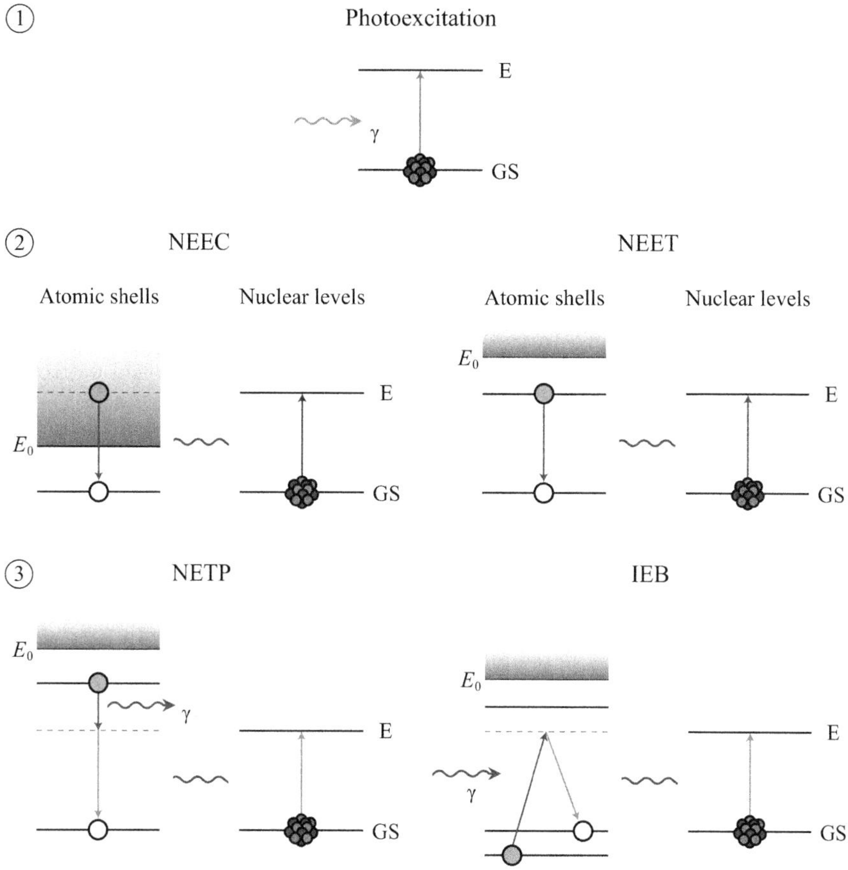

Fig. 2.17 Energy diagrams for several mechanisms of nuclear excitation. The circled numbers indicate the expansion order in perturbation theory. The vacuum level is denoted by E_0, while E and GS represent the excited and ground nuclear states, respectively. The arrowed wavy lines depict real photons, while the non-directional wavy lines depict the exchange of virtual photons. Virtual states are represented by dashed lines

anisms of nuclear excitation and deexcitation are shown in Fig. 2.16 and Fig. 2.17, respectively.

Increasing in complexity, an example of a fourth-order nuclear decay process is the internal conversion in the field of an electron bridge [100]. More complex mechanisms can be identified if the deexcitation involves more than two nuclear states, as in the case of the fourth-order process called electron-nucleus bridge (ENB) [101].

Having reached this point, it seems that the list could go on for a while, and indeed it might, as there are still more mechanisms to be mentioned, such as double-decay processes [95]. Furthermore, the entire framework could be extended to exotic atoms, where the simplest case involves the replacement of the electron with a muon. Therefore, in muonic atoms, one can define the muonic counterpart of the NEET and NEEC processes, namely nuclear excitation upon muonic cascade [102, 103] and NEμC [104], respectively. It is worth mentioning that higher-order does not necessarily imply *tout court* a weaker effect. In fact, in some cases, higher-order decay modes may even become stronger than the first-order decay process. An evident example is provided by the internal conversion, which usually overcomes γ-emission, or by IEB as discussed in Ref. [97].

In the following chapters of this thesis, I mostly focus on the first and second-order electromagnetic mechanisms of nuclear excitation. Chapter 3 provides a detailed presentation of the NEEC process, including an excursus of the ongoing debate surrounding the discrepancy between its experimental observation and theoretical description. This discussion also provides the context in which the study of the NEEC process in excited ions was conceived, with the aim of potentially mitigating this discrepancy. This work culminated in the publication of Ref. [105]. The potential for dynamically controlling nuclear isomer depletion through the use of NEEC and vortex electron beams will also be examined [106]. Chapter 4 continues with the discussion of the muonic counterpart of the NEEC process, which my colleagues and I have named NEμC, which stands for Nuclear Excitation by Free Muon Capture, as published in Ref. [104]. Chapter 5 presents the experimental setup, which I had the opportunity to design and implement at EPFL, that is aimed at the observation of nuclear excitations in optical laser-generated plasma. This chapter also focuses on the theoretical description of the direct photoexcitation process—initiated by the photons produced by the plasma—and on the discussion of the experimental findings.

At this point, I would like to mention a few directions, though not addressed in this thesis, that could be worth investigating in the future. At the moment, suffice it to say that NEμC enables nuclear excitations with energies on the order of ~MeV by providing slow muons, with energies of tens of keV. In contrast to resonant photon absorption, where the energy of the photon must match the energy of the transition, NEμC could offer a more energetically favorable process for isomer depletion or feeding (leaving aside any considerations concerning the muon production). Therefore, it would be interesting to investigate the application of the NEμC process to the depletion schemes of ^{180m}Ta and ^{178m_2}Hf, which are shown in Fig. 2.14 and Fig. 2.10, respectively. Another potential application of NEμC could involve the exploration of E0 transitions (see Table A.2), which have been left out from the calculations of the NEEC probabilities due to their high excitation levels. Given the higher energies at play with NEμC, the excitation of these $I^{\pi} = 0^{+}$ isomeric states from their ground state becomes more feasible, offering valuable insights into the study of rarer decay processes [4]. Even the more standard process of direct photoexcitation appears to still have room for improvement. While the possibility of nuclear excitation by vortex photon beams was proposed in Ref. [107], to my knowledge, there has been no experimental or theoretical investigation of this process.[2] This presents an opportunity to

[2] Since the original thesis was completed, a recent study [108] has theoretically explored collective excitation through vortex photons.

explore high-order multipolar transitions, which—contrary to Table 2.1—could be made more favorable by manipulating the orbital angular momentum carried by the photons, as similarly shown for NEEC in Ref. [106].

In conclusion, much work has yet to be done in order to achieve, if possible, the desired control on the isomer lifetime: one viable path could be to deepen the understanding of atomically-assisted nuclear excitations, mixed perhaps with other interactions. The manipulation of the wavefunction of the particles involved in the process seems to be one such ingredient. In fact, in recent years, vortex beams have shown great potential in nuclear and particle physics, offering new possibilities for the control of the decay modes [106, 109–113].

References

1. Hahn O (1921) Über eine neue radioaktive Substanz im Uran. Berichte der deutschen chemischen Gesel lschaft (A and B Series) 54:1131–1142
2. Fajans K, Morris DF (1973) Discovery and naming of the isotopes of element 91. Nature 244:137–138
3. Shea WR (1983) Otto Hahn and the rise of nuclear physics. Springer Science & Business Media
4. Walker P, Podolyák Z (2020) 100 years of nuclear isomers-then and now. Phys Scr 95:044004
5. Gamow G (1934) Negative protons and nuclear structure. Phys Rev 45:728
6. Jain AK, Maheshwari B, Goel A. (2021) Nuclear isomers: a primer. Springer Nature
7. Audi G, Bersillon O, Blachot J, Wapstra A (1997) The NUBASE evaluation of nuclear and decay properties. Nucl Phys A 624:1–124
8. Audi G, Bersillon O, Blachot J, Wapstra AH (2003) The NUBASE evaluation of nuclear and decay properties. Nucl Phys A 729:3–128
9. Kondev F, Wang M, Huang W, Naimi S, Audi G (2021) The NUBASE2020 evaluation of nuclear physics properties. Chin Phys C 45:030001
10. Garg S et al (2022) Atlas of nuclear isomers-Second edition. Atomic data and nuclear data tables, p. 101546
11. LiveChart of Nuclides (IAEA, Nuclear Data Section). https://nds.iaea.org/ livechart
12. Feenberg E, Hammack KC (1949) Nuclear shell structure. Phys. Rev. 75:1877
13. Feenberg E, Feld BT (1995) Shel l theory of the nucleus. Princeton University Press
14. Mayer MG, Jensen JHD (1960) Elementary theory of nuclear shel l structure. Wiley
15. Woods RD, Saxon DS (1954) Diffuse surface optical model for nucleon-nuclei scattering. Phys Rev 95:577
16. Takigawa N, Washiyama K (2017) Fundamentals of nuclear physics. Springer
17. Lehnert B, Hult M, Lutter G, Zuber K (2017) Search for the decay of nature's rarest isotope Ta180m. Phys Rev C 95:044306
18. (1966) Internal Conversion Processes (ed Hamilton, J.) Academic
19. Hofmann C, Reinhardt J, Greiner W, Schlüter P, Soff G (1990) Angular correlation of electrons and positrons in internal pair conversion. Phys Rev C 42:2632
20. Löbner K (1968) Systematics of absolute transition probabilities of K-forbidden gamma- ray transitions. Phys Lett B 26:369–370
21. Weisskopf V (1951) Radiative transition probabilities in nuclei. Phys Rev 83:1073
22. Blatt JM, Weisskopf VF (1991) Theoretical nuclear physics. Courier Corporation
23. Ring P, Schuck P (2024) The nuclear many-body problem. Springer Science & Business Media
24. Wong SS (2008) Introductory nuclear physics. Wiley

25. Edmonds AR (1957) Angular momentum in quantum mechanics. Princeton University Press
26. Krane KS (1991) Introductory nuclear physics. Wiley
27. Kramp J et al (1987) Nuclear two-photon decay in $0^+ \rightarrow 0^+$ transitions. Nucl Phys A 474:412–450. https://www.sciencedirect.com/science/article/pii/0375947487906257
28. Henderson J et al (2014) Upper Limit on the Two-Photon Emission Branch for the $0_2^+ \rightarrow 0_1^+$ Transition in 98Mo. Phys Rev C 89:064307. https://journals.aps.org/prc/abstract/10.1103/PhysRevC.89.064307
29. De Vries H, De Jager C, De Vries C (1987) Nuclear charge-density-distribution parameters from elastic electron scattering. At Data Nucl Data Tables 36:495–536
30. Tiesinga E, Mohr PJ, Newell DB, Taylor BN (2021) CODATA recommended values of the fundamental physical constants: 2018. J Phys Chem Ref Data 50:033105
31. Walker P, Dracoulis G (1999) Energy traps in atomic nuclei. Nature 399:35–40
32. Segre E (1965) Nuclei and particles: an introduction to nuclear and subnuclear physics. WA Benjamin
33. Weizsäcker V (1936) C. Metastabile Zustände der Atomkerne. Naturwissenschaften 24:813–814
34. Racah G (1952) Nuclear levels and Casimir operator. In: Farkas A, Wigner EP (eds) L. Farkas memorial volume. Research Council of Israel, Jerusalem, pp 294–304
35. Flowers, B. Studies in jj-coupling. I. Classification of nuclear and atomic states. Proceedings of the Royal Society of London. Series A. Mathematical and Physical Sciences 212, 248-263 (1952)
36. Black J, McHarris WC, Kelly W (1971) E6 and M5 Transitions Observed in Fe53m Decay. Phys Rev Lett 26:451
37. Palazzo T et al (2023) Direct Measurement of Hexacontatetrapole, E6 γ Decay from Fe53m. Phys Rev Lett 130:122503
38. Alaga, G., Alder, K., Bohr, A. & Mottelson, B. R. Intensity rules for beta and gamma transitions to nuclear rotational states (1955)
39. Deutsch M, Bauer RW (1960) Spins of the isomeric states of Hf178 and Hf180. Nucl Phys 21:128–132
40. Burson S, Blair K, Keller H, Wexler S (1951) The radiations from hafnium. Physical Review 83:62
41. Bohr A, Mottelson BR (1953) Rotational states in even-even nuclei. Phys Rev 90:717
42. Dracoulis G (2013) Isomers, nuclear structure and spectroscopy. Phys Scr 2013:014015
43. Evaluated Nuclear Structure Data File (NNDC, National Nuclear Data Center) https://www.nndc.bnl.gov/ensdf/. Dec. 2022. (2022)
44. Walker P, Xu F (2015) High-K isomerism in rotational nuclei. Phys Scr 91:013010
45. Helmer R, Reich C (1973) Half-life of 178m2Hf and its neutron capture production. Nucl Phys A 211:1–6
46. Smith M et al (2003) γ rays emitted in the decay of 31-yr 178Hfm2. Phys Rev C 68:031302
47. Walker PM, Dracoulis G (2001) Exotic isomers in deformed atomic nuclei. Hyperfine Interact 135:83–107
48. Eisenberg, J. M. & Greiner, W. Nuclear Theory: Nuclear Models (North-Holland, 1975)
49. Baldwin G, Klaiber G (1947) Photo-fission in heavy elements. Physical Review 71:3
50. Bohr, A. The coupling of nuclear surface oscil lations to the motion of individual nucleons (Munksgaard, 1952)
51. Hill DL, Wheeler JA (1953) Nuclear constitution and the interpretation of fission phenomena. Phys Rev 89:1102
52. Meitner L, Frisch OR (1939) Disintegration of uranium by neutrons: a new type of nuclear reaction. Nature 143:239–240
53. Hahn, O. & Strassmann, F. Concerning the existence of alkaline earth metals resulting from neutron irradiation of uranium. Naturwissenschaften 27 (1939)
54. Polikanov S et al (1962) Spontaneous fission with an anomalously short period. Soviet Physics JETP 15:1016–1021

55. Andreyev A et al (2000) A triplet of differently shaped spin-zero states in the atomic nucleus 186Pb. Nature 405:430–433
56. Runkle R, Champagne A, Engel J (2001) Thermal equilibration of 26Al. Astrophys J 556:970
57. Belic D et al (1999) Photoactivation of 180Tam2 and Its Implications for the Nucleosynthesis of Nature's Rarest Naturally Occurring Isotope. Phys Rev Lett 83:5242
58. Baldwin GC, Solem JC (1997) Recoilless gamma-ray lasers. Rev Mod Phys 69:1085
59. Beeks K et al (2021) The thorium-229 low-energy isomer and the nuclear clock. Nature Reviews Physics 3:238–248
60. Pospelov M, Rajendran S, Ramani H (2020) Metastable nuclear isomers as dark matter accelerators. Physical Review D 101:055001
61. Walker, P. M. & Carroll, J. J. Ups and downs of nuclear isomers. Physics Today 58 (2005)
62. Dracoulis G, Walker P, Kondev F (2016) Review of metastable states in heavy nuclei. Rep Prog Phys 79:076301
63. Aprahamian A, Sun Y (2005) Long live isomer research. Nat Phys 1:81–82
64. Carroll J, Karamian S, Rivlin LA, Zadernovsky A (2001) X-ray-driven gamma emission. Hyperfine Interact 135:3–50
65. Carroll JJ (2007) Nuclear structure and the search for induced energy release from isomers. Nucl Instrum Methods Phys Res, Sect B 261:960–964
66. Sepulveda NA, Jenkins JD, Edington A, Mallapragada DS, Lester RK (2021) The design space for long-duration energy storage in decarbonized power systems. Nat Energy 6:506–516
67. Walker PM, Carroll JJ (2007) Nuclear isomers: Recipes from the Past and Ingredi- ents for the Future. Nucl Phys News 17:11–15
68. Belic D et al (2002) Photo-induced depopulation of the 180Tam2 isomer via low-lying intermediate states: Structure and astrophysical implications. Phys Rev C 65:035801
69. Dracoulis, G. On the question of connections between high-K and low-K states in 180Ta and 176Lu in AIP Conference Proceedings 1269 (2010), 295-302
70. Schlegel C et al (1994) Depopulation of 180Tam by Coulomb excitation and possible astrophysical consequences. Phys Rev C 50:2198
71. Mohr P, Käppeler F, Gallino R (2007) Survival of nature's rarest isotope Ta180 under stellar conditions. Phys Rev C 75:012802
72. Carroll J (2004) An experimental perspective on triggered gamma emission from nuclear isomers. Laser Phys Lett 1:275
73. Collins C et al (1999) Accelerated emission of gamma rays from the 31-yr isomer of 178Hf induced by X-Ray irradiation. Phys Rev Lett 82:695
74. Ahmad I et al (2001) Search for X-Ray Induced Acceleration of the Decay of the 31-Yr Isomer of 178Hf Using Synchrotron Radiation. Phys Rev Lett 87:072503
75. Ahmad I et al (2003) Search for x-ray induced decay of the 31-yr isomer of 178 Hf at low x-ray energies. Phys Rev C 67:041305
76. Carroll J et al (2009) Search for low energy induced depletion of 178Hfm2 at the SPring-8 synchrotron. Phys Lett B 679:203–208
77. Hayes A et al (2006) Breakdown of K Selection in Hf178. Phys Rev Lett 96:042505
78. Roberts H (1997) The importance of stimulated gamma release fromisomers. Hyperfine Interact 107:91–97
79. Ohtsuki T, Yuki H, Muto M, Kasagi J, Ohno K (2004) Enhanced Electron-Capture Decay Rate of 7Be Encapsulated in C60 Cages. Phys Rev Lett 93:112501
80. Cohen R, Miller G, West K (1978) Nuclear resonance excitation by synchrotron radiation. Phys Rev Lett 41:381
81. Pieper G, Heydenburg N (1957) Coulomb excitation of iron-57. Phys Rev 107:1300
82. Petrov YV (1960) Possibility of investigating the levels of the compound nucleus produced by interaction between slow neutrons and isomers. Sov. Phys. JETP 10:833–834
83. Roig O et al (2011) Direct evidence for inelastic neutron "acceleration" by Lu177m. Phys Rev C 83:064617
84. Karamian S, Carroll J (2011) Cross section for inelastic neutron "acceleration" by 178Hfm2. Phys Rev C 83:024604

85. Goldanskii V, Namiot V (1976) Excitation of isomeric nuclear levels by laser radiation via the mechanism of inverse interval electron conversion. Pis'ma Zh Eksp Teor Fiz 23:495
86. Goldanskii, V. & Namiot, V. Excitation of isomeric nuclear levels by laser radiation via the mechanism of inverse internal electron conversion. JETP Lett.(USSR)(Engl. Transl.);(United States) 23 (1976)
87. Dzyublik AY (2013) General theory of nuclear excitation by electron transitions. Phys Rev C 88:054616
88. Carreyre, T. a. et al. First direct proof of internal conversion between bound states. Physical Review C 62, 024311 (2000)
89. Karpeshin F, Trzhaskovskaya M (2017) Bound internal conversion versus nuclear excitation by electron transition: Revision of the theory of optical pumping of the Th229m isomer. Phys Rev C 95:034310
90. Tkalya EV (2003) Probability of L-shell nuclear excitation by electronic transitions in 178Hfm2. Phys Rev C 68:064611
91. Tkalya EV (2005) Induced decay of Hf178m2: Theoretical analysis of experimental results. Phys Rev C 71:024606
92. Dzyublik, A. Y. Triggering of 178Hfm2 by photoinduced electron transition., 11-17 (2013)
93. Kishimoto, S. a. et al. Observation of nuclear excitation by electron transition in 197 Au with synchrotron X rays and an avalanche photodiode. Physical Review Letters 85, 1831 (2000)
94. Krutov V, Fomenko V (1968) Influence of electronic shell on gamma radiation of atomic nuclei. Ann Phys 476:291–302
95. Crasemann B (1973) Some aspects of atomic effects in nuclear transitions
96. Ljubicic A (1991) Rare nuclear processes-investigations with radioisotope sources. J Phys G: Nucl Part Phys 17:S101
97. Berengut J (2018) Resonant electronic-bridge excitation of the U235 nuclear transition in ions with chaotic spectra. Phys Rev Lett 121:253002
98. Volotka A et al (2016) Nuclear excitation by two-photon electron transition. Phys Rev Lett 117:243001
99. Müller RA, Volotka AV, Surzhykov A (2019) Excitation of the Th229 nucleus via a two-photon electronic transition. Phys Rev A 99:042517
100. Krutov V (1990) Internal conversion in the field of an electronic bridge. Sov J Exp Theor Phys Lett 52:584
101. Krutov V, Knyazkov O (1970) Higher approximation effects in internal conversion processes. Ann Phys 480:10–23
102. Wilet L (1954) Excitation of nuclear rotational states In μ-mesonic atoms. Dan: Mat Fys Medd 29
103. Jacobsohn BA (1954) Nuclear fine structure in the μ-mesonic atom. Phys Rev 96:1637
104. Gargiulo S, Gu MF, Carbone F, Madan I (2022) Nuclear excitation by free muon capture. Phys Rev Lett 129:142501
105. Gargiulo S, Madan I, Carbone F (2022) Nuclear excitation by electron capture in excited ions. Phys Rev Lett 128:212502
106. Wu Y, Gargiulo S, Carbone F, Keitel CH, Pálffy A (2022) Dynamical control of nuclear isomer depletion via electron vortex beams. Phys Rev Lett 128:162501
107. Zadernovsky A (2006) Excitation of nuclei by photon beams carrying orbital angular momentum. Laser Phys 16:571–575
108. Lu Z-W et al (2023) Manipulation of giant multipole resonances via vortex γ Photons. arXiv:2306.08377
109. Larocque H, Kaminer I, Grillo V, Boyd RW, Karimi E (2018) Twisting neutrons may reveal their internal structure. Nat Phys 14:1–2
110. Madan I, Vanacore GM, Gargiulo S, LaGrange T, Carbone F (2020) The quantum future of microscopy: wave function engineering of electrons, ions, and nuclei. Appl Phys Lett 116:230502
111. Zhao P, Ivanov IP, Zhang P (2021) Decay of the vortex muon. Phys Rev D 104:036003
112. Luski A et al (2021) Vortex beams of atoms and molecules. Science 373:1105–1109
113. Ivanov IP (2022) Promises and challenges of high-energy vortex states collisions. Prog Part Nucl Phys 103987

Chapter 3
Nuclear Excitation by Electron Capture

It is truly the nature of man to be free and to wish to be so, yet his character is such that he instinctively follows the tendencies that his training gives him. Let us, therefore, admit that all those things to which he is trained and accustomed seem natural to man and that only that is truly native to him which he receives with his primitive, untrained individuality. Thus custom becomes the first reason for voluntary servitude. Men are like handsome race horses who first bite the bit and later like it, and rearing under the saddle a while soon learn to enjoy displaying their harness and prance proudly beneath their trappings.

— Étienne de La Boétie, *Discours de la servitude volontaire*

3.1 From the Origins to the Present Day: A Recipe for the Future

In 1976, Goldanskii and Namiot proposed the possibility of inducing a nuclear excitation through the inverse process of internal electron conversion (IIEC) [1, 2]. Being simply its time reversal, IIEC is a resonant mechanism involving the excitation of the nucleus via the capture of a free electron in an empty shell, provided that the sum of the energy of the free electron E_r and the binding energy E_b of the orbital matches the nuclear transition (E_n), as shown in Fig. 3.1. With rough estimations, the authors calculated a probability of excitation for the IIEC process of $P_{exc} \sim 10^{-14}$ in ^{235}U laser-generated plasma, which was accompanied by a very large resonance cross-section of 10^{-7}b at $E \sim 5$ eV. These values correspond to a rate of $\lambda \sim 10^{-5}s^{-1}$ and an expected number of $N \sim 1000$ excited nuclei per laser pulse for a plasma temperature comparable to the nuclear transition energy $T \sim E_n = 76$ eV. However, the validity of this estimate for the cross-section was considered dubious as it did not take into account the nuclear properties [3].

S. Gargiulo, *Electromagnetic Processes of Nuclear Excitation*, Springer Theses,
https://doi.org/10.1007/978-3-031-56262-4_3

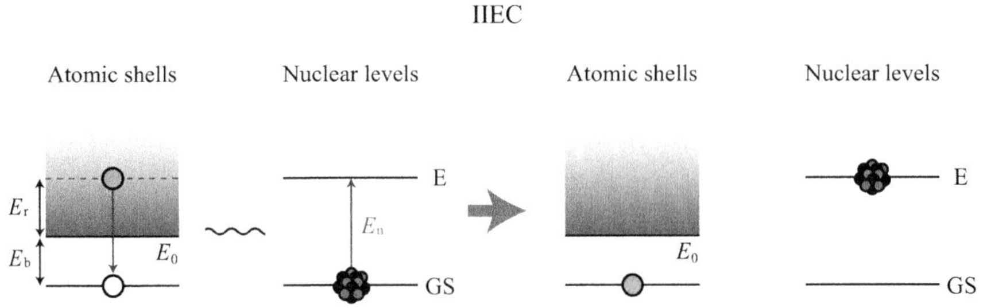

Fig. 3.1 The inverse process of internal electron conversion (IIEC) or, as lately called, nuclear excitation by electron capture (NEEC). Here, a free electron is captured from the continuum in an empty orbital, resonantly exciting the nucleus by exchanging a virtual photon. The excited nucleus (on the right) can either decay by internal conversion or emit a γ-ray

In 1978, Doolen [4, 5] calculated the nuclear deexcitation rate in ^{237}U and ^{238}U. In thermal equilibrium, the rate at which a nuclear level is populated must equal the rate at which it decays. In this way, he obtained the excitation rate for different temperatures and compression factors. He estimated the excitation rate for different processes—including IIEC—and predicted a value of $\lambda^{\text{IIEC}} \sim 10^{-10} - 10^{-8}\text{s}^{-1}$ for the first excited level of ^{237}U at $E = 11.39\,\text{keV}$, assuming a plasma temperature of several keV.

One year later, Izawa and Yamanaka conducted an experiment in a laser-plasma scenario that bore similarities to the one proposed by Goldanskii and Namiot. This experiment resulted in the observation of the first isomeric level of ^{235}U at an energy of 76 eV and half-life of $T_{1/2} = 26$ min. The authors attributed the excitation mechanism leading to the formation of the isomer to the nuclear excitation by electron transition (NEET) process [6]. However, in 1981, Goldanskii and Namiot pointed out that—according to their estimates—the IIEC probability exceeded that of NEET by three orders of magnitude in the experimental conditions of Ref. [6]. Thus, they interpreted Izawa and Yamanaka's observed excitation as due to IIEC [7].

In 1989, Cue et al. proposed the very same IIEC process, which this time did not involve the capture of purely *free* electrons, but rather electrons that were *bound* to atoms or a solid (target material). They named this mechanism nuclear excitation by (target) electron capture (NEEC) [8, 9]. As a consequence of the initial binding state of the captured electrons, the NEEC resonance shape is no longer represented by a Lorentzian function. Instead, it is dominated by the Compton profile of the target electrons, as it will be clear from Fig. 3.5. Cue et al. also drew a connection between NEEC and other known atomic processes that only involved atomic electrons. The most important is the resonant charge transfer and excitation process (RTE), which describes the radiationless capture of a bound target electron into an inner shell of the projectile while simultaneously exciting a projectile electron from a bound state to another level [10–12]. The latter process differs from NEEC in the excitation part: in NEEC it is the nucleus that is excited from one discrete state to another and not an electron as in RTE. Following this logic, the authors provided an order of magnitude estimate for the NEEC cross-section by scaling that of the RTE process. The resulting values are substantially lower than those previously reported by Goldianskii and Namiot and fall in the range $\sim 10^{-3}$ b to 10^{-2} b for several isomers, except for ^{235}U, which was not included in the study. Despite the apparent distinction[1] between the original definitions of the NEEC and IIEC processes, also Cue et al. identified NEEC as the inverse of internal conversion and the two acronyms began to be used interchangeably[2] in the following years.

In 1991, 12 years after the experiment performed by Izawa and Yamanaka, Arutyunyan et al. endeavored to duplicate the same outcomes; however, they failed to observe any excitation of the ^{235}U isomer in a similar laser-plasma scenario [13]. Nevertheless, Arutyunyan et al. were able to successfully detect the $T_{1/2} \sim 26$ min isomer decay by exposing the uranium target to a relativistic electron beam. Regardless of the excitation mechanism, this led them to estimate the excitation cross-section to be between 10^{-8}b and 10^{-7}b. These values were two to three orders of magnitude smaller than those reported in the laser-plasma scenario of Ref. [6], and the unresolved discrepancy remained controversial.

In 1999, Harston and Chemin presented a theoretical overview of the nuclear excitation mechanisms occurring in a plasma environment, which included NEEC [3]. Here, the NEEC cross-section was determined using the principle of detailed balance and the matrix element calculated for its inverse process, i.e. the internal conversion (IC). The excitation rates obtained were on the order of $\lambda \sim 10^{-11}\text{s}^{-1}$ for ^{235}U for plasma temperature between 20 eV–100 eV. Furthermore, in contrast to the estimates reported by Goldanskii and Namiot in Ref. [7], their findings showed that the probability of NEEC to occur in ^{235}U for a plasma temperature of $\sim$100 eV was substantially lower than that of NEET.

In 2003, Mukoyama attempted to clarify the distinction between IIEC and NEEC [14]. In this work, the author introduced a new nomenclature in which the nuclear excitation resulting from the capture of a *free* electron was referred to as NEFEC, while that resulting from a *bound* electron was called NETEC (with the 'T' abbreviating the word 'target'), as depicted in Fig. 3.2. Despite Mukoyama's efforts, this terminology has not been formally adopted afterward, and the acronym NEEC continued to be used interchangeably to describe both processes.

In 2006, Pálffy et al. presented a more sophisticated theoretical framework for NEEC by using a Feshbach projection operator formalism to derive its cross-section. The reported integrated cross-section values for several isomeric transitions ranged from 10^{-5}b eV up to 1 b eV [15, 16]. The theory was subsequently applied to the isomer depletion scheme of Fig. 2.15, predicting NEEC as the most efficient excitation mechanism for $^{242\text{m}}$Am, with a higher cross-section than a previous crude estimate [17], and stronger than direct photoexcitation for $^{93\text{m}}$Mo, $^{178\text{m}}$Hf, and $^{235\text{m}}$U.

---

1 Cue et al. were not aware of the process proposed by Goldianskii and Namiot at the time Ref. [8] was written. However, they later became aware of the earlier proposal, as stated in Ref. [9].

2 Sometimes, the acronym IIC is used in the literature to refer to the inverse process of internal conversion.

Fig. 3.2 Feynman diagrams for the IIEC (NEFEC) and NEEC (NETEC) as proposed by Mukoyama. The diagrams depict the nucleus as a double line, the electron as a thin single line, and the exchange of virtual photons as wavy lines. Dashed lines represent an electron in a bound state, while the asterisk indicates excited states. The labels 'p' and 't' refer to the projectile and target, respectively. The labels 'c' and 'b' stand for continuum and bound, respectively

Although various scenarios have been proposed to observe NEEC experimentally over the years, such as axial channeling [18, 19], electron beam ion traps [15], laser-generated and hot dense plasma [3, 20, 21], and reaction recoil (beam-based) setups [22, 23], the observation of the phenomenon remained elusive for over four decades since its initial proposal.

It was not until 2018 that the first experimental observation of the NEEC process was made in a beam-based scenario on ^{93}Mo [24]. The excitation spectrum of ^{93}Mo includes some of the ingredients that are essential for the isomer depletion scheme depicted in Fig. 2.15, namely: an isomeric state with the energy of 2.425 MeV and half-life $T_{1/2} = 6.85$ h; an upper-lying level located only 4.85 keV away from the isomer, which is an ideal target for depletion, with a half-life of $T_{1/2} = 3.5$ ns; and a ground state with a half-life of $T_{1/2} = 4 \times 10^3$y. In the experiment described in Ref. [24] and illustrated in Fig. 3.3, ^{93}Mo ions were produced through a fusion-evaporation reaction between a 840 MeV ^{90}Zr beam and a ^{7}Li target. This reaction initially populates the energy levels of the ^{93}Mo reaction products ("recoils") that lie above the isomeric state. The subsequent decay of these high energy states may eventually feed the isomer while the ions cross the vacuum gap between the production (^{7}Li) and stopping (^{nat}C) targets. During the passage of the recoils through the stopping target, electrons are initially stripped off, leading to highly charged states in the ^{93m}Mo ions. Later, these electrons recombine back as the recoiling ions are decelerated through collisions with the target atoms. Provided that the energy of these ions is sufficiently high before approaching the stopping medium, the resonance conditions for NEEC (both for energy and charge states) are met as they cross the thick Carbon layer, as depicted in Fig. 3.3a. In this scenario, a prompt coincidence between the γ that feeds the isomer at 2.425 MeV and the sequence of γ-rays with energies of 268 keV, 685 keV, and 1.478 MeV emitted by the stopped nuclei—highlighted in different colors in the level scheme of Fig. 3.3b—would serve as the signature of the isomer

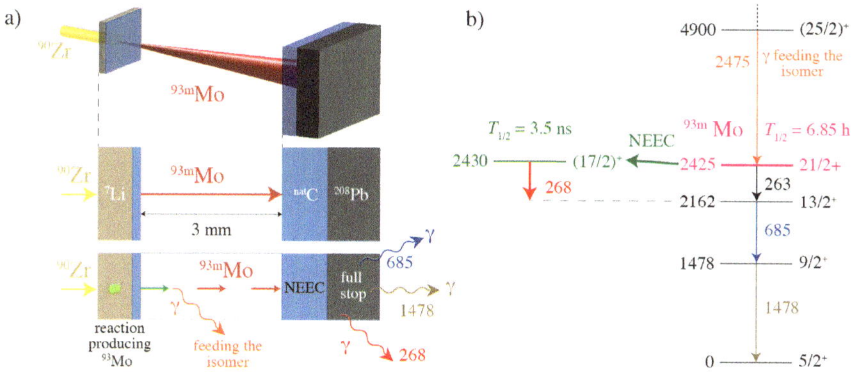

Fig. 3.3 The isomer depletion of ^{93m}Mo as evidence of the process of nuclear excitation by electron capture. a Schematic representation of the experiment performed at the Argonne National Laboratory designed to observe the isomer depletion in ^{93m}Mo. The target composition is shown at the bottom: (i) the fusion-evaporation reaction is induced in the first part of the target—which involves a thin foil of natural Carbon with ^{7}Li evaporated onto it—producing ^{93m}Mo recoils with high velocities; (ii) the vacuum gap of 3 mm between the first and second Carbon layers allows for the isomer to be fed, while ensuring the recoiling ions to meet the resonance once entering the stopping Carbon layer; (iii) as the ions pass through the second ^{nat}C layer, the charge-energy resonance conditions can be met and NEEC might take place; (iv) the C layer is backed by a Pb layer, which completely stops the ions and permits the observation of the γ-decays following the isomer feeding. The natural decay of ^{93m}Mo proceeds with a 263 keV transition, while its depletion is characterized by the 268 keV line. Panel **b** shows the partial energy level scheme of ^{93}Mo. The NEEC process is thought to be responsible for the transition towards the depletion state (in green) lying 4.85 keV above the isomer (in pink). The three γ-rays indicated in orange, blue, and brown can be in true coincidence only if depletion occurs through the red transition. Energies are reported in keV, while the nuclear spin assignments in round parenthesis are uncertain

depletion through a process other than its natural decay with $T_{1/2} = 6.85$ h. As an additional validation, in the absence of depletion, the characteristic 268 keV transition would not appear in the measured spectrum. Chiara et al. observed a depletion of the isomer with a probability of $P_{\text{exc}} \sim 10^{-2}$, and attributed the excitation between the isomeric and depletion states to the NEEC process. This interpretation was driven by the fact that other known processes, such as Coulomb excitation and above-barrier inelastic scattering, which could have played a role in this scenario, have probabilities on the order of $P_{\text{exc}} \sim 10^{-6}$, which are too low to explain the observed phenomenon.

In the same year, further theoretical investigations into the depletion of ^{93m}Mo considered NEEC occurring in an optical laser-generated plasma. These studies detailed the conditions under which the NEEC process could overcome direct photoexcitation from resonant X-rays due to plasma radiation [25, 26]. Later, this scenario was extended by adding an intense X-ray radiation from an X-ray Free Electron Laser (XFEL), which created additional holes in the ions and enhanced the NEEC rate [27]. Still in 2018, a preprint posted on arXiv [28] claimed to have observed the radiative decay of the low-lying ^{229m}Th isomeric state at $E_n \sim 8$ eV in a laser-induced plasma, with the isomer being activated via the NEEC process (referred to as IIC in the paper).

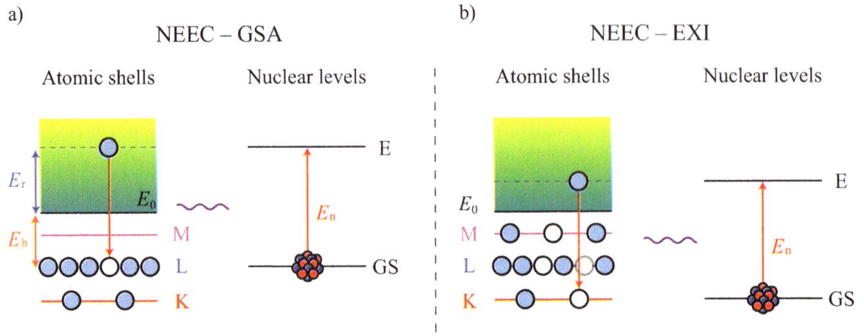

Fig. 3.4 In **a**, the NEEC process occurs in an ion with its atomic shells in the ground state (NEEC–GSA), while in panel **b**, the NEEC process occurs in an ion in an excited electronic state. The inclusion of excited electronic states in the NEEC theory can uncover previously neglected capture channels

However, the complex multi-step process used for the detection and some unclear conditions made it impossible to identify the process unambiguously. To date, these results have not been published in a peer-reviewed journal, as discussed in Ref. [29].

One year later, Wu et al. [30] examined theoretically the conditions of the ^{93m}Mo beam-based experiment performed by Chiara et al.. By considering more than 600 capture channels, they predicted an excitation probability due to NEEC of $P_{exc} \sim 10^{-11}$, which is nine orders of magnitude smaller than the experimental value determined in Ref. [24], and significantly smaller than those resulting from other non-resonant processes.

This significant discrepancy highlighted the necessity of gaining a more comprehensive understanding of the experimental and theoretical frameworks, exploring perhaps aspects that might have been disregarded up to that moment. At the time, it was customary to study NEEC occurring in ground-state electronic configurations, referred to as ground-state assumption (GSA). However, one possible mitigation for the discrepancy between the predicted and experimental excitation probabilities is that NEEC may have taken place in highly excited electronic configurations. In light of these circumstances, in 2020,[3] we proposed a study of the NEEC process occurring in a nucleus with electrons populating atomic orbitals in all possible combinations, as shown in Fig. 3.4, a situation that we identified with the acronym NEEC–EXI [31].

The selection of ^{73}Ge was strategic to clearly demonstrate the potential of NEEC-EXI. This nucleus exhibits a peculiar feature: four electrons filling its orbitals provide sufficient screening to lower the binding energy of the K-shell (E_b^K) below the energy of the isomeric level, which is located at $E_n = 13.28$ keV. Previously, the ground-state assumption precluded NEEC from occurring as a result of the electron capture in the K-shell. However, by removing this hypothesis, our approach revealed the emergence of new capture channels with cross-sections increased by up to three

[3] Although the preprint was made available on arXiv in 2020, it was not officially published until 2022.

orders of magnitude. This enhancement can be understood as follows. By filling the orbitals with electrons, we are passing from a condition in which $E_n < E_b^K$ for $q = Z$—i.e., a bare nucleus for which NEEC in the K-shell is not possible—to a condition where $E_n > E_b^K$ in case of four electrons filling the orbitals (i.e., $q = Z - 4$). The inclusion of excited electronic states in the latter configuration enables NEEC from the capture of electrons in the K-shell and provides orbitals that are nearly resonant with the nuclear transition ($E_b^K \sim E_n$). Since NEEC is a resonant process (i.e., $\sigma_{NEEC} \propto 1/(E_n - E_b^K)$), this circumstance leads to a significant increase in its cross-section. Appendix A.3 offers further details on the recoupling schemes employed for the work presented in Ref. [31].

In 2021, a study by Rzadkiewicz et al. [32] highlighted that the theoretical analysis of the experiment performed at Argonne should have accounted for the fact that the electrons involved in the NEEC process were initially bound in a target medium, rather than purely free as initially assumed by Wu et al. in Ref. [30]. This distinction corresponds to the one previously pointed out by Mukoyama [14] between NEFEC and NETEC, as illustrated in Fig. 3.2. By including the Compton profile of the target electrons in the theoretical analysis, NEEC resonances (here referred to as NEEC–RT to draw a connection to the RTE process) were broadened, resulting in contributions across the entire energy landscape of the projectile ions. This is in contrast to Ref. [30], where resonances appeared at specific discrete energies. Figure 3.5 shows the different cross-section shapes for these two cases. Although the inclusion of the Compton profile made the capture in L orbitals no longer insignificant and comparable to those of other shells, it only slightly increased the NEEC excitation probability to $P_{exc} \sim 7 \times 10^{-11}$, leaving the discrepancy between theory and experiment mostly unaltered. This is because the assumption that atoms are in their electronic ground state before the electron capture suppresses the contribution of the L shells at low projectile energies, where they can actually give most of their contribution. Combining the NEEC-EXI with the NEEC-RT in a future model could further enhance the role of L shells and possibly reduce the disagreement between theory and experiment.[4]

On the experimental side, in 2021, Guo et al. [35] commented on the results of Ref. [24] and attributed the large excitation probability measured in the isomer depletion of ^{93m}Mo to false events, caused by γ-contamination from chance coincidences and Compton background. They estimated the probability of the 268 keV decay to be populated by deexcitation of higher spin states or other reactions and to be in time coincidence (between 100 ns and 1 μs) with any other transitions, without being in true relation to them. The upper limit for this chance probability resulted to be $P_{chance} \sim 5 \times 10^{-3}$, which is close to the experimentally measured NEEC probability ($\sim 10^{-2}$). In addition, they argued that an incorrect background subtraction method could have resulted in the measured excitation probability being only an

[4] Following the completion of the original thesis, a recent study tried to address this aspect [33]. In ^{93m}Mo, the inclusion of excited configurations appears to increase the NEEC probability by a factor of $\simeq 7$ compared to the NEEC-RT predictions. However, it remains unclear to me whether recoupling was considered, and if, given the energy range limitations of ETACHA, there might still be potential for improvements in integrating the NEEC-RT and NEEC-EXI models.

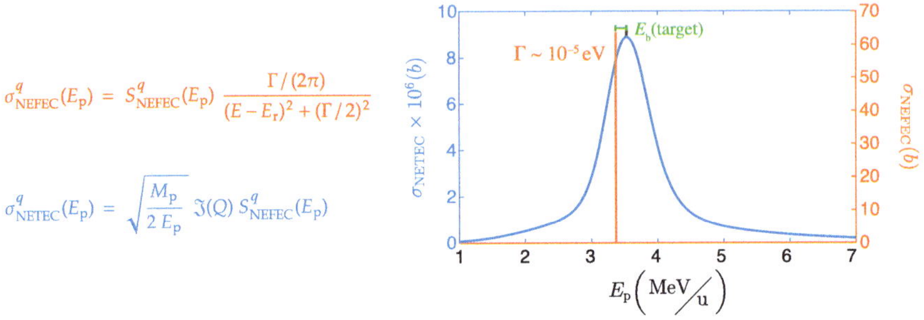

Fig. 3.5 The cross-section shapes for the NEEC process differ depending on whether the electrons involved in the capture are initially bound in a target medium or free. The cross-sections for the NEEC process involving the capture of *free* (NEFEC, in red) or *bound* target electrons (NETEC, in blue) exhibit distinct differences in their shapes as a function of the energy of the projectile ions (E_P). NEFEC cross-section has a Lorentzian shape, with a width typically in the range $\gamma \sim 10^{-10} - 10^{-5}$eV, while the NETEC cross-section extends over a much broader energy range, due to the Compton profile of the target electrons $\mathfrak{J}(Q)$. The distance between the two peaks is due to the fact that, in NETEC, the captured electron is bound in the target medium. In the figure, this distance is represented by E_b(target) $\sim$ 100 eV. The NETEC cross-section presented in this figure was calculated for the 4.85 keV transition in ^{93}Mo, utilizing the Compton profile from Ref. [34] and the model described in Refs. [11, 32]. In this example, the resonance strength S_{NEFEC} was chosen to be 10^{-3}b eV

upper limit for the isomer depletion of $^{93\mathrm{m}}$Mo. To address these concerns, Guo et al. suggested repeating the measurements with the $^{93\mathrm{m}}$Mo production site several meters away from the stopping target. By doing so, prompt γ-rays due to the production of ^{93}Mo and relative deexcitations from high spin states would be separated from the γ-transitions due to the isomer depletion.

A response to this comment was given by Chiara et al. in the same issue [36]. They followed the same steps as Guo et al. to calculate the chance probability but pointed out some differences in the values used for the estimation. By considering a smaller time window of 90 ns and a lower rate of 1.7 kHz for the population of the 268 keV transition from higher states, Chiara et al. concluded that $P_{\mathrm{chance}} \sim 3 \times 10^{-4}$, which is smaller than the observed depletion. As a side note, it should be mentioned that this value corresponds to the lower end of the range quoted by Guo et al. for P_{chance}. To validate their calculations, Chiara et al. also extracted the chance-coincidence probability from experimental data, which resulted in a comparable value of $P^{\mathrm{exp}}_{\mathrm{chance}} \sim 8 \times 10^{-4}$. They also argued that statistical errors arising from these chance coincidences could lower the measured value of P_{exc} to $8.8 \times 10^{-3} - 9.6 \times 10^{-3}$. In this reply, the authors provided additional details about the background subtraction and discussed the outcomes of their analysis in contrast to the points raised by Guo et al.. In addition, Chiara et al. investigated the effect of varying the background parameter within a certain range on the calculated value of P_{exc}. Their analysis revealed only a minor change of approximately 9% in the final value of the

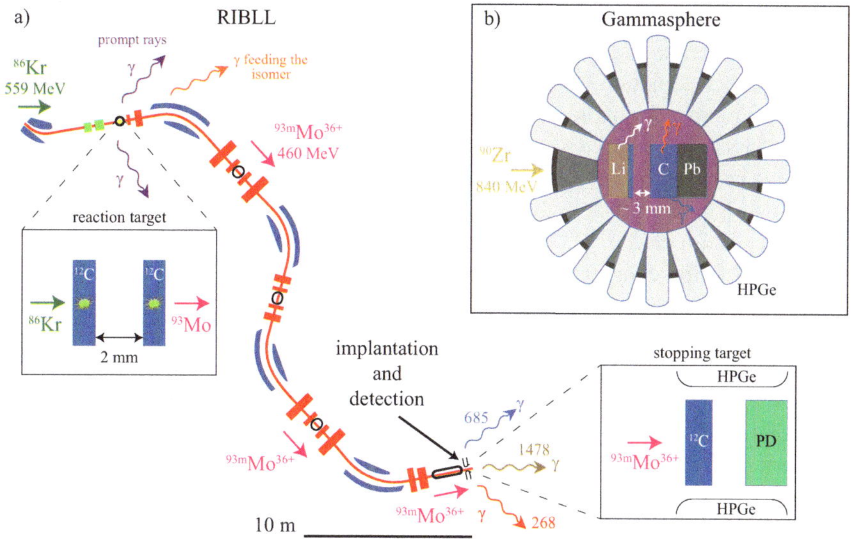

Fig. 3.6 Reinvestigation of the isomer depletion occurring in ^{93}Mo at RIBLL. In **a** the Radioactive Ion Beam Line in Lanzhou RIBLL used by Guo et al.. Highly charged ^{93}Mo^{36+} recoils are selected following their production by fusion-evaporation ^{12}C(^{86}Kr,5n) reaction at the primary target. This secondary beam is then delivered to the stopping medium, consisting of a Carbon foil and a plastic detector (PD), located 35 m away from the reaction target. In this configuration, prompt and delayed (originating from the isomer) γ-rays are spatially separated, leading to a low γ-background in the detection system. Implantation events are recorded by the plastic detector (PD), while the γ-transitions are detected by high purity germanium (HPGe) detectors. For comparison, panel **b** shows the configuration used by Chiara et al. in the Gammasphere (ANL), where both the primary and secondary targets are surrounded by the detection system

excitation probability. Therefore, in their opinion, the experimental case is robust and NEEC cannot be excluded as the mechanism responsible for the observed isomer depletion.

In 2022, Guo et al. [37] attempted to reproduce the beam-based experiment on ^{93m}Mo at the Radioactive Ion Beam Line in Lanzhou (RIBLL) [38], shown in Fig. 3.6a. The main focus of this repetition was on the spatial separation between the prompt γ-rays arising from the fusion-evaporation reaction and the deexcitation of high spin states of ^{93}Mo, and those—delayed—originating from the isomer depletion. In the experimental setup shown in Fig. 3.6a, the reaction target produces ^{93}Mo recoils through the reaction ^{12}C(^{86}Kr, 5n), and is located 35 m away from the stopping target consisting of a carbon foil and a plastic detector (PD). NEEC can occur in the stopping target where the detection system is placed. In contrast, the configuration used by Chiara et al. [24] (Fig. 3.6b) had both the reaction target and stopping medium placed inside the Gammasphere spectrometer [39]. Therefore, this reinvestigation under low γ-background aimed at a higher sensitivity measurement of the isomer depletion occurring in ^{93m}Mo.

The ^{93}Mo recoils are delivered to the stopping medium 1.14 μs after their production at the reaction target, providing sufficient time for the isomer to be fed. In the case of isomer depletion, a sequence of γ-decays would be detected in coincidence (within a few nanoseconds) with the implantation events recorded in the PD. In the experiment conducted by Guo et al., the 268 keV γ-ray transition was not observed, leading to an experimental upper limit for the isomer depletion of $P_{exc}^{max} \sim 10^{-5}$. This upper limit is approximately three orders of magnitude smaller than the value determined by Chiara et al. [24]. Theoretical calculations presented for the experimental conditions of Guo et al. predicted a probability of isomer depletion due to NEEC of $P_{exc} \sim 10^{-12}$. These calculations included the momentum distribution of the target electrons but once again assumed the electronic shell to be in the ground state prior to the electron capture (NEEC–GSA).

Although the kinetic energies of the ^{93m}Mo recoils were different between the two experiments (674 MeV for Chiara et al. and 460 MeV for Guo et al.), the large discrepancy in the measured P_{exc} values seems difficult to explain solely by the different experimental conditions that can be learned from the respective papers. It is worth noting that the different energies in turn affect the average charge states of the ions during the slowdown process in the stopping medium. It is clear that a new attempt with a low γ-background and higher recoil energies is desirable to further investigate the discrepancy in the measured excitation probabilities. On the other hand, exploring other types of experimental scenarios is also necessary to provide a comprehensive test of the theory. While the internal conversion can indirectly prove—due to time reversal—the NEEC theory for the capture of free electrons, an equivalent observation for solid-state targets seems to be lacking.

In 2022, we presented in Ref. [40] (by Wu et al.) a novel approach for the enhancement of the NEEC cross-section through the manipulation of the quantum state (i.e., the wavefunction) of the captured electrons. This concept is illustrated in Fig. 3.7. This work represents a significant milestone, in my opinion, as it challenges the conventional vision of nuclear reactions from here on. Traditionally, in most textbooks and literature, the cross-section of such processes is often attributed to the unique properties of the nucleus and the kinematics of the "projectiles", as probably may have also appeared to the reader from earlier discussions. In the case of NEEC, for instance, the overlap between the wavefunctions of bound and continuum states is a crucial factor in determining the final rate of occurrence. However, since *free* particles are usually treated as plane waves, the NEEC cross-section for a specific orbital and ionization state was previously thought to be entirely determined by the nucleus, almost as if it was an inherent property of nature. In Ref. [40], it was theoretically shown for the first time that the cross-sections of atomically assisted nuclear excitation processes, such as NEEC, may not be solely determined by nuclear properties but can be influenced by engineering the wave function of the electron undergoing capture, as conceptually illustrated in Fig. 3.7. Specifically, nuclear excitation by vortex electron capture was investigated for the isomer depletion schemes of ^{93m}Mo, shown in Fig. 3.3, and of ^{152m}Eu, shown in Fig. 3.8a. The use of a vortex beam with a topological charge $m = 5$ for ^{93m}Mo leads to a two-orders-of-magnitude increase

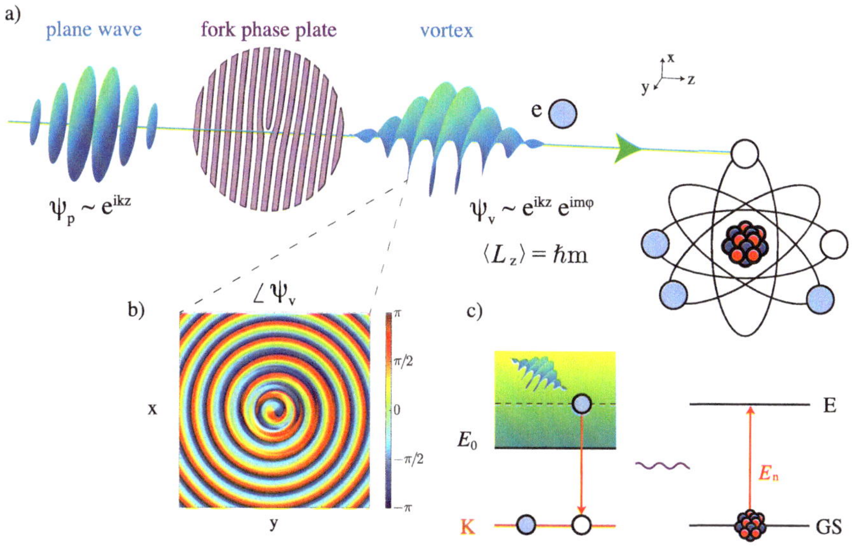

Fig. 3.7 Nuclear excitation by vortex electron capture. In **a**, the electron wavefunction, initially represented as a plane wave ψ_p, undergoes manipulation through the interaction with a fork phase plate. This imparts a well-defined orbital angular momentum (OAM) $\hbar m$ to the electrons along the axis of propagation ($\langle L_z \rangle = \hbar m$). In panel **b** is shown an example of the phase ($\angle\psi_v$) that can be imprinted on the electron wavefunction by using phase plates or plasmonic structures, as demonstrated in Refs. [41, 42]. Panel **c** illustrates the energy diagram of the NEEC process with an incoming free electron in a vortex state. The energy-matching conditions still need to be satisfied

over the strongest capture channel of the plane wave case. However, an even more intriguing aspect arises for the depletion of $^{152\mathrm{m}}$Eu.

The use of vortices on $^{152\mathrm{m}}$Eu has the striking feature of changing the shell for which the cross-section is strongest, as depicted in Fig. 3.8b. In the case of the plane waves, capture in the $2s_{1/2}$ shell is the most probable. As mentioned earlier, the shell for which the capture and simultaneous nuclear excitation is most efficient is entirely determined by the nucleus, selection rules of the transition, and the overlap between the continuum and bound state wavefunctions (which are again determined by the nucleus and its ionization state). The use of vortex beams breaks this traditional paradigm and reveals that the efficiency of nuclear excitation can depend on specific manipulations of the electron wavefunction. Specifically, a vortex beam with $m = 3$ externally selects the strongest capture channel to be the $2p_{1/2}$ orbital (as indicated by the pink arrow labeled 'plane to vortex' in Fig. 3.8b). By changing the topological charge to $m = 5$, the excitation through the $2p_{3/2}$ orbital becomes the most efficient channel, as shown by the pink arrow labeled 'OAM change'. Therefore, the manipulation of the orbital angular momentum (OAM) of the incoming electron beam, which can be achieved on an ultrafast timescale with plasmonic structures [41] or using light phase masks [44, 45], allows for external control over the cross-section of the process. Furthermore, when physical phase plates are employed, as the fork

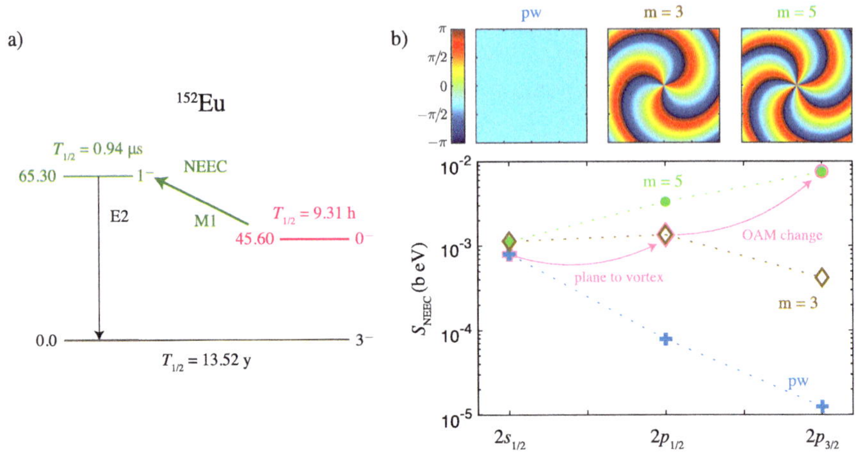

Fig. 3.8 **Isomer depletion of ^{152m}Eu through vortex electron beams.** In **a**, the energy level scheme for the isomer depletion of ^{152m}Eu is shown. The isomeric state is drawn in pink, while the depletion state is in green. The energies are given in keV. In **b**, the resonance strengths for NEEC are compared for plane waves (blue) and vortex electron beams with topological charges $m = 3$ (brown) and $m = 5$ (green). The phase of the electron wavefunction is shown for the plane wave (PW) and vortex beams (Laguerre-Gaussian solutions [43]) with $m = 3$ and $m = 5$. The markers outlined in pink indicate the strongest capture channel for a given beam solution or topological charge. The pink arrows illustrate the shift of the strongest capture channel towards higher-ℓ subshells as we move from plane wave solutions to vortex states

phase plate shown in Fig. 3.7, deflector coils can be utilized to choose the desired vortex topological charge [46]. This newfound ability to control the cross-section through the manipulation of the electron wavefunction brings us closer to realizing the long-sought dream of electromagnetically manipulating the isomer decay. Using NEEC as a testbed, it is clear that applying these concepts to other nuclear excitation mechanisms can lead to unexpected outcomes and perhaps lay the groundwork for incorporating additional aspects to make these processes even more efficient.

From a historical perspective, it is evident that the NEEC process remains a conundrum even almost half a century after its proposition. Recent publications have contributed to improving our understanding of the experimental and theoretical frameworks in which NEEC can occur, incorporating ingredients from the past—that sometimes went lost—and looking towards the future for the implementation and application of the newly developed concepts.

Having reviewed the literature in the preceding pages, it is clear that the superimposition of different acronyms to refer to the same process has led to a perception of chaos (to which I have also partially contributed). My preferred terminology for the process of nuclear excitation by electron capture is NEEC, regardless of whether the capture involves a *free* or *bound* (target) electron. However, I believe it is necessary to specify from time to time the scenario under study in order to avoid confusion. This is also the approach I took in the next chapter, where the process of nuclear excitation by free muon capture is presented. In this case, I used the acronym NEμC to refer to the process and specified in the text that the capture involves a free particle, although it would have been less plausible to have target muons. As Latins would say: *"de gustibus non est disputandum"*.

References

1. Goldanskii V, Namiot V (1976) Excitation of isomeric nuclear levels by laser radiation via the mechanism of inverse internal electron conversion. JETP Lett.(USSR)(Engl. Transl.);(United States) 23
2. Goldanskii V, Namiot V (1976) Excitation of isomeric nuclear levels by laser radiation via the mechanism of inverse interval electron conversion. Pis'ma Zh Eksp Teor Fiz 23:495
3. Harston M, Chemin J (1999) Mechanisms of nuclear excitation in plasmas. Phys Rev C 59:2462
4. Doolen G (1978) Nuclear excitation rate. Phys Rev Lett 40:1695
5. Doolen G (1978) The nuclear excitation rate for low-lying excited states in U237 and U238. Phys Rev C 18:2547
6. Izawa Y, Yamanaka C (1979) Production of 235Um2 by nuclear excitation by electron transition in a laser produced uranium plasma. Phys Lett B 88:59–61
7. Goldanskii V, Namiot V (1981) Excitation of nuclear isomeric levels in a hot plasma by inverse internal electron conversion. Sov J Nucl Phys (Engl. Transl.) 33
8. Cue N, Poizat J-C, Remillieux J (1989) Exciting the nucleus by target electron capture into atomic orbitals. EPL (Europhys Lett) 8:19
9. Cue N (1989) Nuclear excitation by target electron capture. Nucl Instrum Methods Phys Res, Sect B 40:25–27
10. Tanis J et al (1981) Simultaneous electron capture and excitation in S + Ar Collisions. Phys Rev Lett 47:828
11. Brandt D (1983) Resonant transfer and excitation in ion-atom collisions. Phys Rev A 27:1314
12. Mokler P, Reusch S (1988) Comments on correlated electron capture in relativistic, highly charged, heavy ions. Zeitschrift für Physik D Atoms, Molecules and Clusters 8:393–395
13. Arutyunyan R et al (1991) Cross section for excitation of the isomer 235mU in the plasma produced by an electron beam. Sov J Nucl Phys (English Translation) 53:23–26
14. Mukoyama T (2003) In: Nuclear excitation involving inner-shel l electrons in AIP conference proceedings, vol 652, pp 281–290
15. Pálffy A (2006) Theory of nuclear excitation by electron capture for heavy ions https://core.ac.uk/download/pdf/56343138.pdf. PhD thesis. Justus-Liebig- Universität. arXiv: 0701206 [physics]
16. Pálffy A, Scheid W, Harman Z (2006) Theory of nuclear excitation by electron capture for heavy ions. Phys Rev A 73:012715
17. Zadernovsky A, Carroll J (2002) Non-radiative triggering of long-lived nuclear isomers. Hyperfine Interact 143:153–174
18. Dauvergne D (2006) Expériences interdisciplinaires d'interaction particule-matière PhD thesis
19. Morel P, Daugas JM, Gosselin G, Gogny D (2005) Nuclear excitation by electronic processes: NEEC and NEET effects plasma at local, pp 1085-1088
20. Gosselin G, Morel P (2004) Enhanced nuclear level decay in hot dense plasmas. Phys Rev C 70:064603
21. Gosselin G, Méot V, Morel P (2007) Modified nuclear level lifetime in hot dense plasmas. Phys Rev C 76:044611
22. Karamian S, Carroll J (2012) Calculated yield of isomer depletion due to NEEC for 93mMo recoils. Phys At Nucl 75:1362–1367
23. Polasik M et al (2017) Resonance conditions for Mo 93 m isomer depletion via nuclear excitation by electron capture in a beam-based scenario. Phys Rev C 95:034312

24. Chiara C et al (2018) Isomer depletion as experimental evidence of nuclear excitation by electron capture. Nature 554:216–218
25. Wu Y, Gunst J, Keitel CH, Pálffy A (2018) Tailoring laser-generated plasmas for efficient nuclear excitation by electron capture. Phys Rev Lett 120:052504
26. Gunst J, Wu Y, Keitel CH, Pálffy A (2018) Nuclear excitation by electron capture in optical-laser-generated plasmas. Phys Rev E 97:063205
27. Wu Y, Keitel CH, Pálffy A (2019) X-ray-assisted nuclear excitation by electron capture in optical laser-generated plasmas. Phys Rev A 100:063420
28. Borisyuk P et al (2018) Excitation of 229Th nuclei in laser plasma: the energy and half-life of the low-lying isomeric state. arXiv:1804.00299
29. Thirolf P, Seiferle B, Von der Wense L (2019) The 229-thorium isomer: doorway to the road from the atomic clock to the nuclear clock. J Phys B: At Mol Opt Phys 52:203001
30. Wu Y, Keitel CH, Pálffy A (2019) Mo^{93m} isomer depletion via beam-based nuclear excitation by electron capture. Phys Rev Lett 122:212501
31. Gargiulo S, Madan I, Carbone F (2022) Nuclear excitation by electron capture in excited ions. Phys Rev Lett 128:212502
32. Rzadkiewicz J, Polasik M, Słabkowska K, Carroll J, Chiara C et al (2021) Novel approach to Mo93m isomer depletion: nuclear excitation by electron capture in resonant transfer process. Phys Rev Lett 127:042501
33. Rzadkiewicz J, Słabkowska K, Polasik M, Carroll J, Chiara C et al (2023) Mo 93 m isomer depletion via nuclear excitation by electron capture in resonant transfer into highly excited open-shell atomic states. Phys Rev C 108:L031302
34. Biggs F, Mendelsohn L, Mann J (1975) Hartree-Fock Compton profiles for the elements. At Data Nucl Data Tables 16:201–309
35. Guo S, Fang Y, Zhou X, Petrache C (2021) Possible overestimation of isomer depletion due to contamination. Nature 594:E1–E2
36. Chiara C et al (2021) Reply to: possible overestimation of isomer depletion due to contamination. Nature 594:E3–E4
37. Guo S et al (2022) Probing Mo93m isomer depletion with an isomer beam. Phys Rev Lett 128:242502
38. Sun Z, Zhan W-L, Guo Z-Y, Xiao G, Li J-X (2003) RIBLL, the radioactive ion beam line in Lanzhou. Nucl Instrum Methods Phys Res, Sect A 503:496–503
39. Lee I-Y (1990) The gammasphere. Nucl Phys A 520:c641–c655
40. Wu Y, Gargiulo S, Carbone F, Keitel CH, Pálffy A (2022) Dynamical control of nuclear isomer depletion via electron vortex beams. Phys Rev Lett 128:162501
41. Vanacore GM et al (2019) Ultrafast generation and control of an electron vortex beam via chiral plasmonic near fields. Nat Mater 18:573–579
42. Rotunno E et al (2021) One-dimensional ghost imaging with an electron microscope: a route towards ghost imaging with inelastically scattered electrons. arXiv:2106.08955
43. Bliokh KY et al (2017) Theory and applications of free-electron vortex states. Phys Rep 690:1–70
44. Lembessis V, Ellinas D, Babiker M, Al-Dossary O (2014) Atom vortex beams. Phys Rev A 89:053616
45. Lembessis VE (2017) Atomic Ferris wheel beams. Phys Rev A 96:013622
46. Pohl D et al (2017) Atom size electron vortex beams with selectable orbital angular momentum. Sci Rep 7:934

Chapter 4
Nuclear Excitation by Free Muon Capture

Our systems of post-experimental thought and our systems of motivation and action therefore co-exist in paradoxical union. One is "up-to-date"; the other, archaic. One is scientific; the other, traditional, even superstitious. We have become atheistic in our description, but remain evidently religious—that is, moral—in our disposition. What we accept as true and how we act are no longer commensurate. We carry on as if our experience has meaning—as if our activities have transcendent value—but we are unable to justify this belief intellectually. We have become trapped by our own capacity for abstraction: it provides us with accurate descriptive information but also undermines our belief in the utility and meaning of existence.

— Jordan Peterson, *Maps of Meaning*

4.1 Why Muons?

Several ingredients have to come together for the NEEC process to occur and to be efficient: (i) the presence of vacancies in which the capture can occur; (ii) a good match between atomic binding energies and nuclear transition energies (i.e., $E_{\mathrm{b}} \sim E_{\mathrm{n}}$); and (iii) a significant overlap between bound and continuum wavefunctions. While the first (i) and third (iii) requirements can be considered somewhat accessible, the range of nuclear levels efficiently reachable through NEEC is limited by the K-shell binding energy of a given nuclide. This is due to the inverse proportionality of the NEEC cross-section with the energy of the free particle that has to be captured, that is $\sigma_{\mathrm{NEEC}} \propto {}^{1}/_{(E_{\mathrm{n}}-E_{\mathrm{b}})}$. Considering that $E_{\mathrm{b}}^{\mathrm{K}}$ is approximately 110 keV for $Z = 90$, it becomes increasingly less probable to excite nuclear levels substantially higher in energy due to the growing energy mismatch. Moreover, 100 keV nuclear transitions could already be very challenging to achieve for much lighter atoms.

S. Gargiulo, *Electromagnetic Processes of Nuclear Excitation*, Springer Theses,
https://doi.org/10.1007/978-3-031-56262-4_4

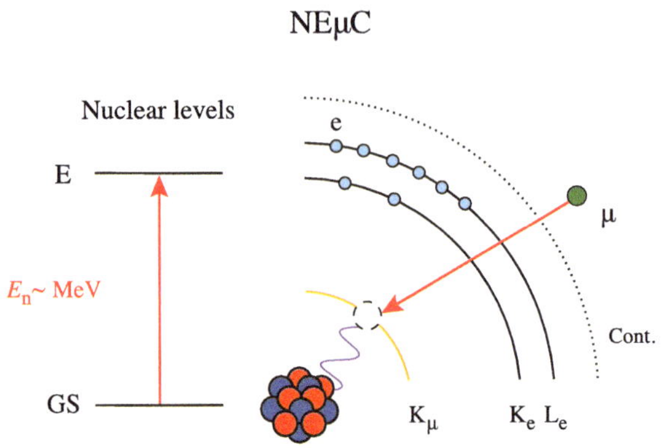

Fig. 4.1 The process of nuclear excitation by free muon capture (NEμC). The figure shows the muonic (yellow) and electronic (black) orbitals. Unlike electronic orbitals, muonic shells are always free for capture in atoms. As a result, ionization requirements are no longer necessary, breaking another paradigm of the NEEC process

One of Italy's most renowned philosophers and poets, Giacomo Leopardi, regarded Nature as an evil and indifferent step-mother [1], creator of a need for pleasure that can never be satisfied, depriving human beings of the possibility of achieving happiness. Often, one can see the goal, but Nature does not provide an opportunity to reach it. Going beyond human perceptions of happiness in relation to their personal narratives [2–5], this philosophical thought is highly present in various aspects of physics: Nature presents us with trade-offs with which we must contend and often leaves us a little leeway, demanding greater effort to overcome them. The idea of a process equivalent to NEEC, occurring in exotic atoms, emerged from similar patterns of thought. If shells closer to the nucleus were available, they would have had higher binding energies: (i) making $\sim$ MeV nuclear levels easier to excite and (ii) enhancing the interaction with the nucleus due to the increased proximity. However, within the realm of conventional atoms, possibilities are pretty much determined. As a result, our attention has turned to muonic atoms. Following this line of thinking, in 2022, we proposed the nuclear excitation by free muon capture (NEμC) process [6], depicted in Fig. 4.1. Using muons has a first immediate advantage: the ionization requirement is no longer needed because muonic orbitals are always available for capture due to the instability of this subatomic particle.

A comparison between NEEC and its muonic counterpart (NEμC) is presented in Fig. 4.2. In principle, both processes can excite the same nuclear transition with an energy of $E_n \sim$ MeV. However, the roles of the resonance E_r and binding E_b energies are reversed between the two processes, while they add up to E_n. In conventional atoms, the binding energy E_b is generally smaller than 110 keV, which means that the remaining energy must be provided in the form of kinetic energy of the free particle to fulfill NEEC resonance conditions, as shown in Fig. 4.2a. This situation is reversed in the case of muonic atoms [7–9]. The binding energy for the K muonic shell is on the order of ~ 12 MeV for $Z = 90$ and up to ~ 17 MeV for superheavy elements. As seen in Fig. 2.5, the highest isomeric energy state occurs in ^{208m}Pb at 13.67 MeV. Therefore, in muonic atoms, it is much more likely to find situations where orbitals are almost resonant with such high-energy states, potentially covering the entire range of isomeric energies. In principle, NEμC might be used to address

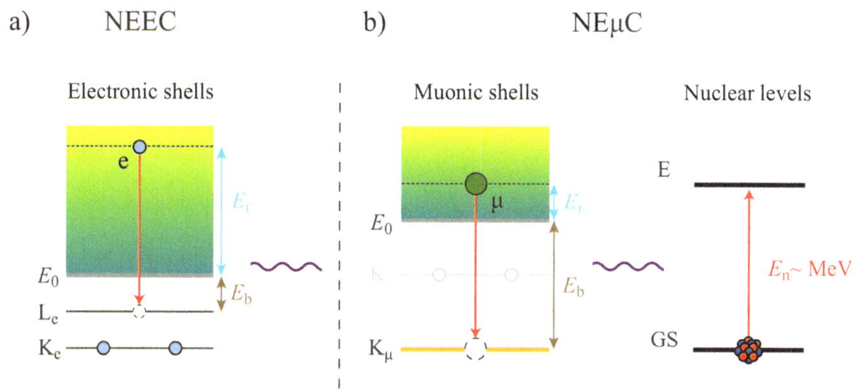

Fig. 4.2 Comparison between the NEEC and NEμC processes. The energy diagrams for the NEEC (**a**) and NEμC (**b**) processes are displayed. The roles of the binding (E_b) and resonance (E_r) energies are reversed between the two mechanisms while adding up to reach a nuclear excitation in the MeV range ($E_n = E_r + E_b$). Electronic shells are indicated with the label 'e' while muonic ones with a 'μ'. E_0 represents the vacuum level. In (**b**), the electronic shells are shaded to improve readability

both the depletion of an isomer inducing a transition towards an intermediate level–as seen in Fig. 2.14 for ^{180m}Ta, where a ~ 1 MeV excitation is needed to release the 77 keV stored in the long-lived isomer–as well as the activation of high-energy states, initiating the transition from the ground state.

In Ref. [6], we proposed NEμC by examining all nuclear excited levels–regardless of their lifetime–connected to the ground state by low-order electric multipolar transitions, since these are generally dominant as discussed in Chap. 2. To perform the necessary calculations, the Dirac equations had to be solved for the Coulomb potential in the presence of muons [10, 11]. At that time, the most commonly used freely-available computational packages for atomic calculations, such as GRASP2K [12] and FAC [13], did not have the capability to handle muonic atoms, despite the similarity with the electronic case and the framework used therein. As part of the collaboration that arose for the realization of the work presented in Ref. [6], the Flexible Atomic Code (FAC) was updated with state-of-the-art theoretical models to include muons as well. These modifications made the FAC the only freely-available online tool that can solve the Dirac equations for muonic atoms within 1 keV accuracy when compared to other state-of-the-art theoretical calculations [14]. As anticipated from our initial considerations, the integrated cross-section for NEμC was found to be up to ten orders of magnitude larger than that of NEEC and two to five orders of magnitude higher than that of direct photoexcitation for ~ MeV quadrupolar transitions. This implies that in certain cases, NEμC is the most efficient mechanism for exciting such nuclear levels, especially considering the absence of a tunable source of high-intensity MeV photons. It is worth noting, however, that the transitions identified in Ref. [6] do not lead–directly–to the population of isomers due to their selection criteria. The majority of these excited levels have half-lives $T_{1/2} \ll 1$ ns, as shown

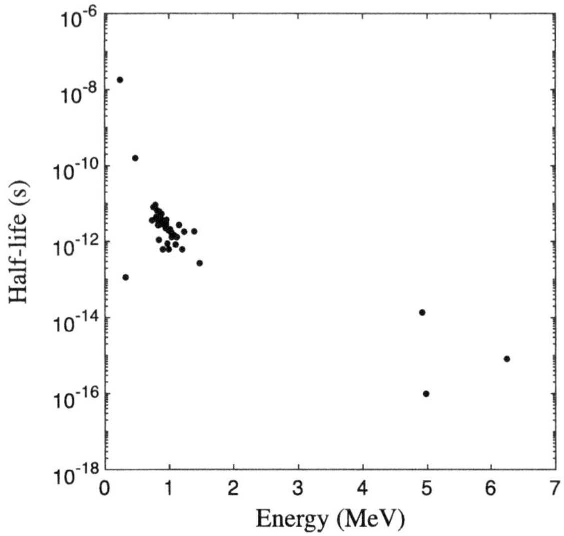

Fig. 4.3 Half-life and energy map of the nuclear levels excited through NEμC, following the selection criteria of Ref. [6] for E1 and E2 transitions

in Fig. 4.3. Nevertheless, direct isomer feeding from the ground state is possible if different criteria are considered.

The unique properties of NEμC call for its application to the isomer depletion scheme depicted in Fig. 2.15, where a long-lived isomer is connected to a fast-decaying depletion level through a low-order multipolar transition, followed by the release of energy via γ-ray emission or a converted electron. Identifying levels located above the isomer that can act as a depletion state is crucial, as these may have no direct transition to the isomer or their reduced transition probability $B(\lambda L)$ may be unknown. A possible step forward in the foreseeable future might be the identification of such levels in the excitation spectra of ^{178m}Hf (shown in Fig. 2.10) and ^{180m}Ta (shown in Fig. 2.14) that could enable an efficient isomer depletion through NEμC. In ^{180m}Ta, NEμC could provide an alternative method for depleting the 77.2 keV isomer by employing muons with much lower energies than those required by the photons for direct photoexcitation, which are typically on the order of $\sim$ MeV [15–17].

Considering the high energy at play, it was reasonable to expect that NEμC could have had a significant impact on inducing prompt fission of the nucleus by providing sufficient energy to overcome the barrier, which is $E_{\mathrm{f}} \sim 6$ MeV for actinides. It is well known that when muons slow down in a target medium, they are typically captured in high principal quantum number orbitals (with $n \sim 14$). This initiates a cascade of multiple transitions that ends with the muon reaching the K-shell within 100 fs [7, 18]. During the cascade, muons can induce prompt nuclear fission by exciting the giant dipole (GDR) or quadrupole (GQR) resonances, transferring energy to the nucleus through radiationless transitions between two bound muonic states [18–20]. This excitation mechanism, shown in Fig. 4.4a, can be seen as the muonic analogous of the NEET process for electrons and could be called NEμT, but it is more commonly referred to as nuclear excitation by muonic cascade. However, more likely, the muon

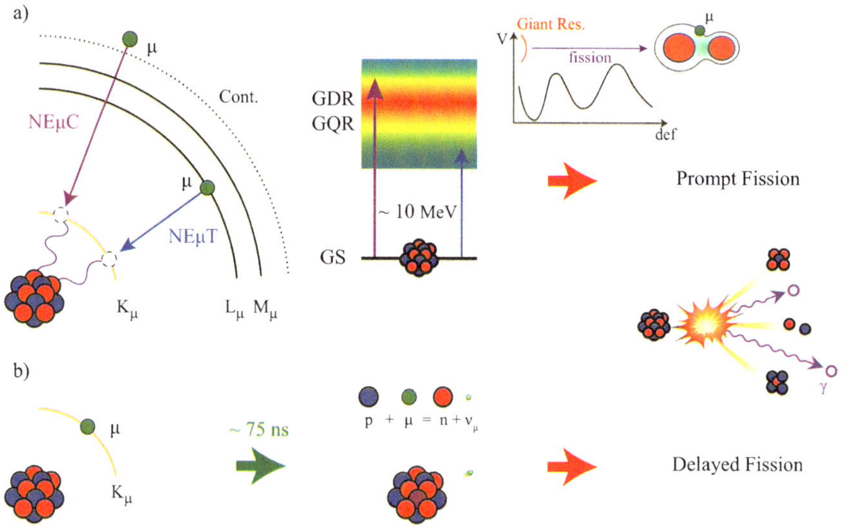

Fig. 4.4 Overview of the muon-induced fission processes. In panel (**a**), radiationless transitions from free to bound states (NEμC) or between two bound states (NEμT) can provide enough energy to the nucleus to excite the giant dipole (GDR) or quadrupole (GQR) resonances. These states can act as a doorway to prompt fission having enough energy to overcome the double-humped barrier. In panel (**b**), the muon can be captured by a proton while it sits in the K-shell, leading to the reaction $(Z, A) + \mu = (Z-1, A)^* + \nu_\mu$ in a timescale of $\tau_{cap} \sim 75$ ns for actinides [21, 22]. The resulting fission is therefore delayed compared to the fragments produced by the processes described in (**a**) by the lifetime of the weak decay process (τ_{cap})

will cascade down to the K-shell by emission of Auger electrons ($n > 5$) or through radiative X-rays ($n \leq 5$). At this point, the muon can be captured by a proton in the nucleus while it sits in the K-shell, resulting in the formation of a neutron and a muon neutrino [18], as shown in Fig. 4.4b. The average excitation energy provided to the nucleus is generally on the order of 15 MeV, which is sufficient to overcome the fission barrier for actinides $E_f \sim 6$ MeV. The fission resulting from nuclear muon capture is delayed compared to the moment the muon entered the atom and the fragments produced by NEμT, by the characteristic time of the weak decay process, which is $\tau_{capt} \sim 75$ ns for actinides [21, 22].

Our work highlights a third channel, the fission induced by the radiationless capture of a free muon in a bound orbital (NEμC), which seems not to have been taken into account in previous studies to explain the experimental observations [23–26]. This process is depicted in Fig. 4.4a, together with its competition with NEμT. Therefore, we calculated the probability of inducing fission by NEμC through the excitation of the GDR in ^{238}U by capturing a free muon into the K-shell [6]. This probability was found to be equal to $P_f \sim 4 \times 10^{-5}$ per incident muon. However, it should be noted that this value only accounts for the K-shell and does not include contributions from the excitation of the GQR. Having these aspects in mind, the probability of fission induced by NEμC appears to be significantly weaker than the prompt fission induced by NEμT (with $P_f \sim 10^{-3} - 10^{-2}$) and the delayed fission resulting from nuclear muon capture ($P_f \sim 10^{-2} - 10^{-1}$) [24]. Despite the results of our calculations, fission induced by NEμC–occurring in shells higher than K–does not preclude the possibility of detecting a subsequent muonic X-ray[1] in coincidence with the fission event. Therefore, it is possible that the contribution of NEμC to prompt fission has been incorrectly attributed and merged with the one caused by NEμT in experimental observations that rely on double (γf) or triple coincidences ($\mu^- \gamma f$) events as a signature. Further investigations are desirable, particularly in light of the controversial experimental results discussed in Refs. [18, 27]. It is worth noting that when moving towards lighter isotopes, the innermost transitions (E1: $2p - 1s$; E2: $3d - 1s$) could eventually lose their overlap with the giant resonances, resulting in a weaker NEμT–induced fission. In this case, the additional degree of freedom provided by NEμC, in the form of the kinetic energy of the free lepton, may allow it to become the dominant process for prompt fission. However, a specific study is required to determine whether this is possible or not.

As mentioned in Chap. 2, the NEμC process could also be useful in the study of E0 monopole transitions, which are reported in Table A.2. These transitions cannot be excited by single-photon absorption, and NEμC may prove to be an efficient way to reach these low-lying 0^+ isomers directly from the ground state. Once excited, these 0^+ states cannot decay via a single gamma-ray emission, making higher-order processes such as internal conversion and pair production crucial for their deexcitation. If, as mentioned by Phil Walker in Ref. [28], one considers the scenario where all electrons have been removed, and the transition energy E_n is less than 1022 keV, then both channels of internal conversion and pair production are suppressed. In such cases, it should be possible to observe rarer decay modes. As an additional element to this recipe, it may be of interest to consider the excitation of the E0 transition through NEμC and the subsequent presence of a bound muon in the K-shell. Under specific conditions, such as when the muon capture rate is much smaller than the muon decay rate ($\lambda_{capt} \ll \lambda_\mu$) and the isomer has a half-life in the presence of the bound muon (i.e., including the internal conversion of the bound muon) $T_{1/2} \gg {}^1/_{\lambda_\mu}$, I can think of a special case, portrayed in Fig. 4.5, though perhaps quite impractical to study in reality. As the decay of a bound muon in a bound electron is a rare event [29, 30], the muon is likely to decay by emitting a free electron, leaving the isomer completely bare and with suppressed internal conversion (IC) channels, for both muons and electrons. If this sequence occurs, it could be possible to excite the 0^+ levels through NEμC from the ground state and have the opportunity to study rare decay modes afterward. However, several conditions must be met for this pattern to occur, and identifying a favorable excitation state is crucial.

Recently, I became aware of an earlier attempt by Kaufmann and Pilkuhn in 1977 to calculate the probability of inducing neutron emission in ^{208}Pb through free-bound muon transitions, which corresponds to what we have referred to as NEμC,

---

[1] The term "muonic X-ray" refers to a photon emitted as a result of the transition of the muon between two states.

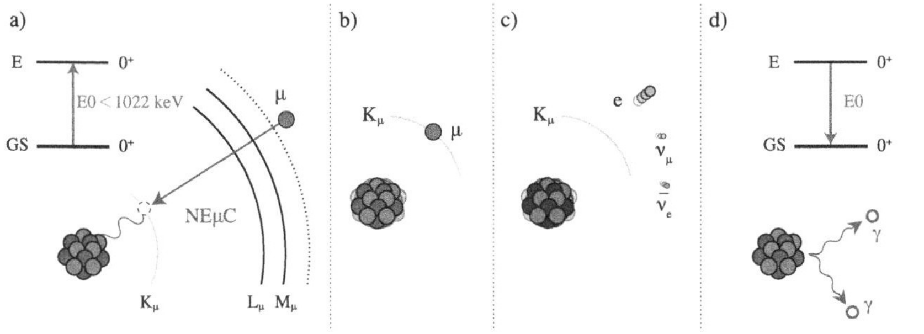

Fig. 4.5 Possible reaction pattern to study rare decay modes through monopole transitions: in panel (**a**), the E0 transition is induced by the capture of a free muon (NEμC), connecting the ground state to a low-lying 0^+ isomer. (**b**) Thereafter, the nucleus will be in an excited state and the muon will be bound to the K-shell. In (**c**), the muon decays resulting in the emission of a free electron. In panel (**d**), the bare atom cannot decay by single-photon emission, internal conversion, or internal pair formation, leaving room for rare decay processes such as the two-photon decay mode. Between panels (**b**) and (**c**), it is possible for the nucleus to deexcite through internal conversion of the bound muon or to undergo nuclear muon capture

using a non-relativistic surface transition model [31]. It is surprising to learn that the theoretical backbone of the nuclear excitation through the radiationless capture of a free lepton was presented[2] just one year after the initial proposition of the NEEC process by Goldanskii and Namiot [32, 33]. Nonetheless, as discussed in Chap. 3, it took nearly 30 years for NEEC to have a more rigorous derivation [34, 35]. Curiously, subsequent works and reviews on the topic [7, 9, 20, 36, 37] seem to have lost trace of this process, and no direct connection with NEEC was ever drawn. In fact, the paper by Kaufmann and Pilkuhn has received only two citations in 46 years and, in both cases, it is mentioned among "mechanisms that proceed via electron capture or scattering" [38, 39].

It is not uncommon for scientific discoveries to be made through a convoluted pattern of research efforts. On this aspect, it was extremely instructive for me to learn–from a talk by Prof. Barry Marshall–about the tortuous path that led to the discovery and oblivion, repetitively over a century, of the Helicobacter pylori [40–42].

This feels to me as an eternal return [43].

References

1. Leopardi, G. La Ginestra (1836). Available with parallel italian and english texts on http://www.tclt.org.uk/leopardi/Canti_2011.pdf, page 212

[2] It appears that Kaufmann and Pilkuhn were not aware of the NEEC process.

2. Csikszentmihalyi M, Nakamura J, Csikszentmihalyi M (2014) The concept of flow. In: Flow and the foundations of positive psychology: the collected works of Mihaly Csikszentmihalyi.
3. Peterson JB (2002) Maps of meaning: the architecture of belief. Routledge
4. Zagrebelsky G (2015) Liberi servi. Il Grande Inquisitore e l'enigma del potere, Torino: Einaudi
5. Freud S (1929) Civilization and its discontents
6. Gargiulo S, Gu MF, Carbone F, Madan I (2022) Nuclear excitation by free muon capture. Phys Rev Lett 129:142501
7. Devons S, Duerdoth I (1969) Muonic atoms. Adv Nucl Phys 2:295–423
8. Borie E, Rinker G (1982) The energy levels of muonic atoms. Rev Mod Phys 54:67
9. Nagamine K (2003) Introductory muon science. Cambridge University Press
10. Nikiforov AF, Novikov VG, Uvarov VB, Uvarov V (2005) Quantum-statistical models of hot dense matter: methods for computation opacity and equation of state. Springer Science & Business Media
11. Grant IP (2007) Relativistic quantum theory of atoms and molecules: theory and computation. Springer
12. Jönsson P, Gaigalas G, Bieroń J, Fischer CF, Grant I (2013) New version: Grasp2K relativistic atomic structure package. Comput Phys Commun 184:2197–2203
13. Gu MF (2008) The flexible atomic code. Can J Phys 86:675–689
14. Michel N, Oreshkina NS, Keitel CH (2017) Theoretical prediction of the fine and hyperfine structure of heavy muonic atoms. Phys Rev A 96:032510
15. Belic D et al (1999) Photoactivation of ^{180}Ta^{m2} and its implications for the Nucleosyn-thesis of nature's rarest naturally occurring isotope. Phys Rev Lett 83:5242
16. Belic D et al (2002) Photo-induced depopulation of the ^{180}Ta^{m2} isomer via low-lying intermediate states: Structure and astrophysical implications. Phys Rev C 65:035801
17. Walker PM, Carroll JJ (2007) Nuclear isomers: recipes from the past and Ingredi-ents for the future. Nucl Phys News 17:11–15
18. Oberacker V et al (1993) Muon-induced fission: a probe for nuclear dissipation and fission dynamics. Phys Rev C 48:1297
19. Wheeler JA (1949) Some consequences of the electromagnetic interaction between μ-mesons and nuclei. Rev Mod Phys 21:133
20. Zaretski D, Novikov V (1961) Theory of nuclear excitations by muons in heavy mesic atoms. Nucl Phys 28:177–191
21. Goulard B, Primakoff H (1974) Nuclear muon-capture sum rules and mean nuclear excitation energies. Phys Rev C 10:2034
22. Schröder W et al (1979) Evidence for atomic muon capture by fragments from prompt fission of muonic Np^{237}, Pu^{239}, and Pu^{242}. Phys Rev Lett 43:672
23. Belovitskii G et al (1960) On the mechanism of uranium fission induced by slow negative muons. Sov Phys JETP 11:296–299
24. Ahmad S et al (1980) Fission yields and lifetimes for muon induced fission in ^{235}U and ^{238}U. Phys Lett B 92:83–86
25. Johansson T et al (1980) Muon induced quadrupole photofission. Phys Lett B 97:29–32
26. Hänscheid H et al (1992) Nuclear excitation and prompt fission in muonic ^{238}U. Zeitschrift für Physik A Hadrons and Nuclei 342:111–120
27. Oberacker VE, Umar AS, Karpeshin FF (2004) Prompt muon-induced fission: a sensitive probe for nuclear energy dissipation and fission dynamics. arXiv:nucl-th/0403087
28. Walker P, Podolyák Z (2020) 100 years of nuclear isomers-then and now. Phys Scr 95:044004
29. Greub C, Wyler D, Brodsky S, Munger C (1995) Atomic alchemy: weak decays of muonic and pionic atoms into other atoms. Phys Rev D 52:4028
30. Aslam MJ, Czarnecki A, Zhang G, Morozova A (2020) Decay of a bound muon into a bound electron. Phys Rev D 102:073001
31. Kaufmann W, Pilkuhn H (1977) Nuclear excitation via free-bound muon atomic transitions. Zeitschrift für Physik A Atoms and Nuclei 280:283–285
32. Goldanskii V, Namiot V (1976) Excitation of isomeric nuclear levels by laser radiation via the mechanism of inverse interval electron conversion. Pis'ma Zh Eksp Teor Fiz 23:495

33. Goldanskii V, Namiot V (1976) Excitation of isomeric nuclear levels by laser radiation via the mechanism of inverse internal electron conversion. JETP Lett.(USSR)(Engl. Transl.);(United States) 23
34. Pálffy A, Scheid W, Harman Z (2006) Theory of nuclear excitation by electron capture for heavy ions. Phys Rev A 73:012715
35. Pálffy A (2006) Theory of nuclear excitation by electron capture for heavy ions. PhD thesis, Justus-Liebig-Universität. https://core.ac.uk/download/pdf/56343138.pdf. arXiv: 0701206 [physics]
36. Measday DF (2001) The nuclear physics of muon capture. Phys Rep 354:243–409
37. Hughes V (2012) Muon physics: electromagnetic interactions. Elsevier
38. Shahbaz A (2009) Muonic atoms in super-intense laser fields
39. Shahbaz A, Müller C, Bürvenich TJ, Keitel CH (2009) Laser-induced nonresonant nuclear excitation in muonic atoms. Nucl Phys A 821:106–117
40. Warren JR, Marshall B (1983) Unidentified curved bacilli on gastric epithelium in active chronic gastritis. Lancet 321:1273–1275
41. Marshall B, Warren JR (1984) Unidentified curved bacilli in the stomach of patients with gastritis and peptic ulceration. Lancet 323:1311–1315
42. Egan BJ, O'Morain CA (2007) A historical perspective of Helicobacter gastroduo-denitis and its complications. Best Pract & Res Clin Gastroenterol 21:335–346
43. Nietzsche F (2008) Thus spoke Zarathustra: a book for everyone and nobody. Oxford University Press

Chapter 5
Nuclear Excitations in Optical-Laser Generated Plasma

Listen, Alëša", Ivan began in a resolute voice, "if I am really able to care for the sticky little leaves I shall only love them, remembering you. It's enough for me that you are somewhere here, and I shan't lose my desire for life yet. Is that enough for you? Take it as a declaration of love if you like..

The Brothers Karamazov, — Fëdor Dostoevskij

The advent of the high-intensity laser made it possible to study a variety of phenomena bridging nuclear and atomic physics [1, 2], as those seen in Chap. 2. Not only can the laser serve as a tool to alter the occurrence of these mechanisms, but it can also create the stage where these processes can enact. One possibility is the formation of a large concentration of free ions and electrons, formally a neutral plasma or cluster [3, 4], trough ablation [5, 6] by shining high-intensity laser on nanoparticles or macroscopic solid targets. Plasma fulfills many of the conditions necessary for these processes to occur, at once. The simultaneous presence of free electrons, nuclei, vacancies, and broad spectra of X-ray photons, makes the study of low-energy nuclear excitations very appealing in this environment. This unique mixture allows for the coexistence of two or more processes competing for the excitation of nuclear levels, with the dominant one determined by plasma conditions and the specific isotope being studied. As a result, plasma could become the theater of simultaneous nuclear excitations due to direct photon absorption, NEEC [7–10], NEET [7, 11], inverse electron bridge [12] and inelastic scattering of electrons with nuclei [2, 13, 14].

The study of nuclear excitation processes in a plasma scenario is of paramount significance in the determination of the population of long-lived excited nuclei in astrophysical plasma, as they can have a substantial impact on the nucleosynthesis pathways in stellar locations [15–17]. A long-standing dream was also the possibility of achieving population inversion between paired nuclear levels in stationary

S. Gargiulo, *Electromagnetic Processes of Nuclear Excitation*, Springer Theses,
https://doi.org/10.1007/978-3-031-56262-4_5

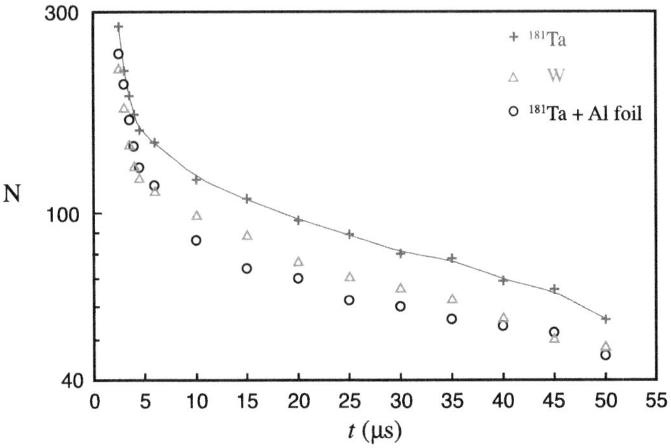

Fig. 5.1 Tantalum and tungsten deexcitation in laser-generated plasma experiments. Total number (N) of detected events as a function of time t for ^{181}Ta (+), W (△) and ^{181}Ta+Al foil (∘) targets. The figure has been redrawn and adapted from Ref. [24]

plasma, which was particularly captivating for the pursuit of γ-ray lasers [18–20]. However, despite numerous theoretical proposals, the experimental observation of many processes of nuclear excitation in the plasma environment is currently either lacking or controversial. One example, as discussed in detail in Chap. 3, is the experiment performed by Izawa and Yamanaka on laser-generated ^{235}U plasma. Initially, the observed excitation was attributed to NEET [21] and later suggested to be due to NEEC [22]. However, a later repetition of the experiment produced no supporting evidence of the isomer excitation [23].

Even more conventional processes, such as direct photoexcitation, may still lack clear evidence in this chaotic scenario. In 1999, Andreev et al. [24] reported the observation of nuclear excitation of the low-lying ^{181}Ta isomer (6.237 keV) in an expanding plasma generated by a femtosecond laser ($\lambda = 600$ nm and $\tau_{\text{pulse}} = 200$ fs). Their results have been redrawn and reported in Fig. 5.1, where the number of events as a function of time is shown. According to their calculations, direct photoexcitation was the dominant process leading to this nuclear transition [25]. In Ref. [24], the ^{181}Ta signal was measured and compared to the signal originating from a tungsten target (W), which leads to comparable plasma properties, although the two nuclei have significantly different nuclear levels and half-lives (approximately 6 μs for ^{181}Ta and approximately 1 ns for W). The excess of 49 events in Ta compared to W was considered a sign of the achieved nuclear excitation. Nevertheless, the time dependence of both Ta and W signals are very similar and essentially overlap in case an aluminum foil is added as a filter. This could suggest that the time response is primarily dominated by the residual afterglow of the scintillator or by the recombination of the ions on the chamber walls (acting in the interval $1 - 100$ μs as stated in Ref. [24]). Since even subtle changes in surface properties or the presence of impurities can significantly affect the ablation process, it is desirable to obtain a clearer

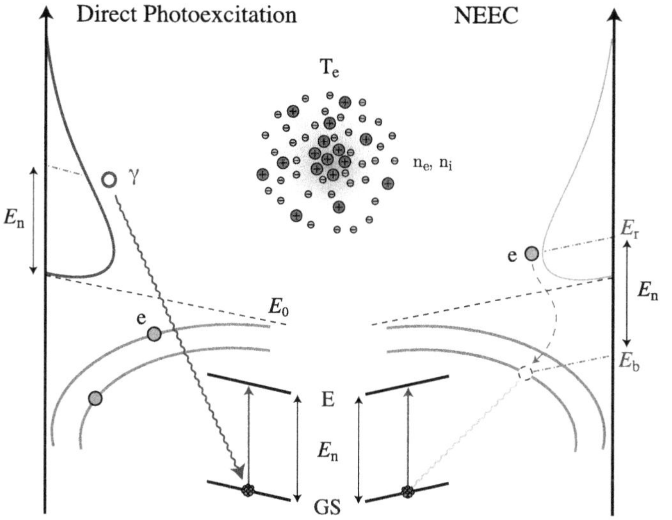

Fig. 5.2 Competition between direct photoexcitation through plasma resonant X-ray photons and NEEC in optical-laser generated plasma. Excitation through NEEC is possible only if vacancies are present. Here, the kinetic energy of the free electron, able to induce the nuclear transition E_n through radiationless electron capture in a shell with binding energy E_b, has to satisfy the resonance condition $E_r = E_n - E_b$. T_e, n_e, and n_i synthetically symbolize the plasma properties, which are the electron temperature, and electron and ion densities, respectively. The vacuum level E_0 is represented by the dashed line. This image has been inspired by Fig. 1 of Ref. [26]

experimental observation that does not rely on excess counts. Specifically, it is hoped that a significant difference in the decay rates of the signals obtained with Ta and W targets could be observed.

Experimental challenges in the identification of the mechanisms leading to the nuclear transition are usually related to the fine-tuning of the plasma parameters and the possibility of having a sort of *switch* that could clearly enable one process over another. In many cases, a single mechanism dominates and leads to the excitation of the nuclei, making the identification clear. However, in other cases, this isolation is not possible, and different processes may overlap and compete in intensity. This makes the theoretical description of the plasma crucial for the correct interpretation. An example of such a case is the competition between direct excitation through plasma thermal radiation and NEEC, as depicted in Fig. 5.2. This has been studied in detail for the isomer depletion of ^{93m}Mo in Refs. [9, 10].

While the process of direct photoexcitation (DP) can take place independently from the charge state of the ions, NEEC requires the presence of vacant shells where the free electrons can be captured. A detailed description of the charge states of ions and populations of electronic levels becomes fundamental since, as seen in Chap. 3, NEEC is strongly influenced by the atomic surrounding. If we consider an expanding plasma in contrast to a confined one, the description of these properties becomes time-dependent. Currently, available particle-in-cell (PIC) simulations, such as EPOCH [27] and VLPL [28], or radiative-collisional codes as FLYCHK [29], only provide the average charge state of the ions during the expansion. Also for this reason, NEEC in expanding plasma has been extensively studied, assuming the electronic configuration to be in its ground state prior to the electron capture. A more detailed description of the properties of the plasma and their influence on the nuclear excitation rate would provide a more comprehensive understanding and potentially reveal interesting effects that have yet to be explored, as seen in the case of NEEC-EXI [30].

In this chapter, I will present the design and implementation of a table-top setup for studying nuclear excitations in expanding plasma generated by a femtosecond optical laser. Finally, I will discuss the outcomes of the experimental measurements.

Regardless of the excitation process, the sequence of events depicted in Fig. 5.3 can be imagined to occur. In principle, various pathways are available for the exci-

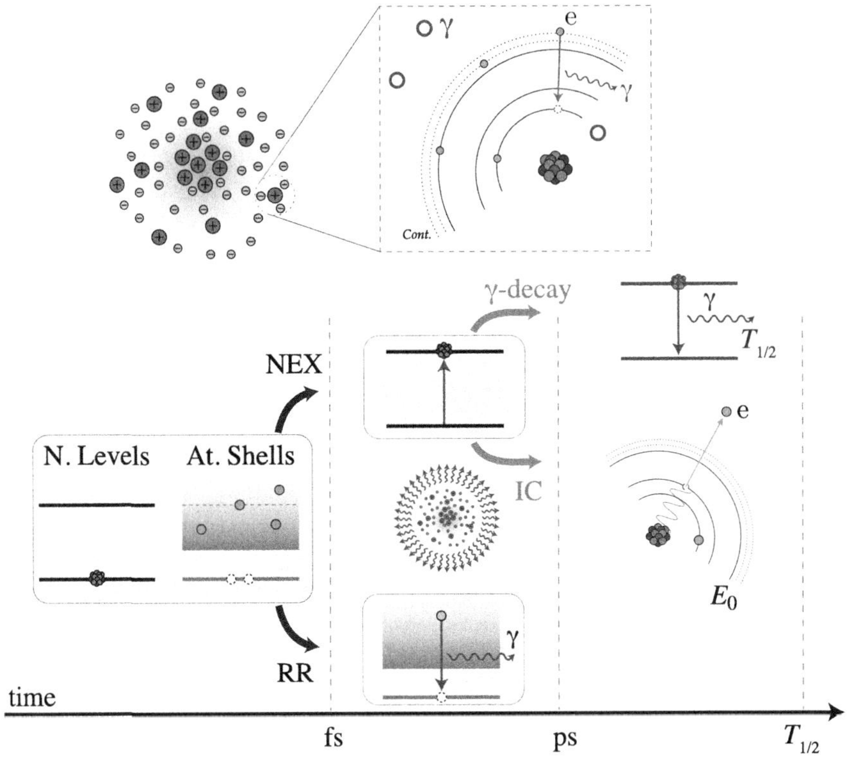

Fig. 5.3 Temporal discrimination between plasma radiation and delayed nuclear decay. After laser ablation, atoms are in highly ionized states, while their nuclei are in the ground state. In this environment, the coexistence of photon and electron distributions can lead to nuclear excitation (NEX) through various mechanisms, such as direct photoexcitation, NEEC, inelastic scattering, NEET, and others. Regardless of the excitation process, the subsequent decay by γ-ray emission or IC will be delayed by the half-life $T_{1/2}$ of the excited level. Thus, X-rays coming from the plasma radiation and γ-rays due to nuclear decay can be discriminated in the time domain

tation of nuclei in the presence of a population of photons and free electrons. These pathways can mainly be categorized as direct absorption of photons, atomically-assisted processes, or inelastic scattering with free electrons. Once the nucleus has been excited, it will generally decay by internal conversion (IC) or γ-emission with a characteristic time given by the half-life $T_{1/2}$. If an isomer is excited from the fundamental state, its subsequent nuclear decay would be delayed by at least several ns with respect to the background signal caused by plasma radiation and radiative recombinations (RR), which generally persist throughout the plasma lifetime (i.e., up to hundreds of ps). Although the different processes overlap in energy, their distinct timescales allow for the identification of nuclear decay through time-resolved spectroscopic techniques.

5.1 Search of Suitable Isotopes

As a first step, I performed an isotope scan using data from the IAEA's NDS LiveChart of Nuclides database [31]. The scan was based on criteria that favored isotopes with stable or long-lived ground states, a first excited state of a few tens of keV, and a half-life of the excited state higher than several tens of ps. These criteria were chosen for experimental feasibility and safety reasons, as well as to ensure detectability with commercially available X-ray detectors. More precisely, the following criteria were used:

- Transitions involving a nuclear excitation from the ground state;
- Stable or long-lived ground state ($T_{1/2}^{\mathrm{GS}} \geq 1 \times 10^3$ y);
- $E_\mathrm{n} \leq 100\,\mathrm{keV}$;
- Half-life of the excited state $T_{1/2}^{\mathrm{E}} \geq 100\,\mathrm{ps}$.

The outcomes of this investigation highlighted 80 isotopes and 507 nuclear transitions, shown in Fig. 5.4 with respect to the nuclear transition energy and in Fig. 5.5 with respect to the nuclear level half-life.

Figure 5.4 also includes the binding energies of K, L, M, and N shells in a bare nucleus—calculated using the Flexible Atomic Code (FAC) [32]—to aid in the selection of isotopes suitable for the NEEC process. As discussed in Chap. 3, exploring other experimental scenarios in which NEEC can take place could be invaluable in improving our understanding of this phenomenon, especially given the current controversy surrounding it. NEEC occurs only in shells with energy levels E_b below the nuclear transition energy E_n. The difference between these energies is known as the resonance energy E_r of the continuum free electron that must be captured. Since NEEC is a resonant process, its cross-section is proportional to $\sigma_{\mathrm{NEEC}} \propto {}^1/_{(E_\mathrm{n}-E_\mathrm{b})}$, making it most efficient when there is a lower mismatch between these energies. The half-life ($T_{1/2} = {}^{\hbar \ln 2}/_{\Gamma_\mathrm{n}}$) reported in Fig. 5.5 includes all decay channels, i.e. $\Gamma_\mathrm{n} = (1+\alpha_{\mathrm{IC}})\Gamma_\gamma$, where Γ_n is the level width, Γ_γ is the width due to the γ-decay channel and α_{IC} is the internal conversion coefficient for a neutral atom [33].

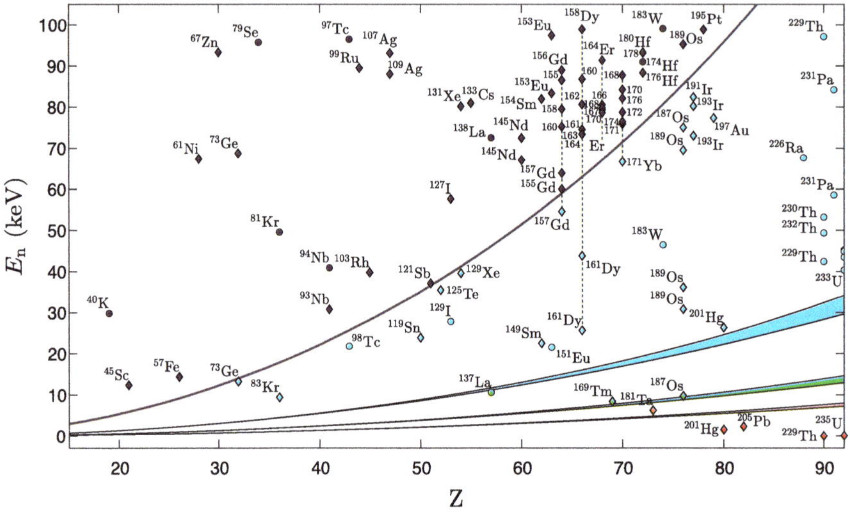

Fig. 5.4 Nuclear transition energies of the isotopes that meet the search criteria. Excitation energies as a function of the atomic number Z for isotopes falling in the search criteria. The K (dark purple), L (blue), M (green), and N (orange) shell binding energies for bare nuclei have been included as colored lines. The vertical dashed lines group elements with the same atomic number. Diamond markers indicate stable ground states, and their color refers to the first inner shell for which NEEC is possible. The four isotopes noted with red markers are below the N shell binding energy

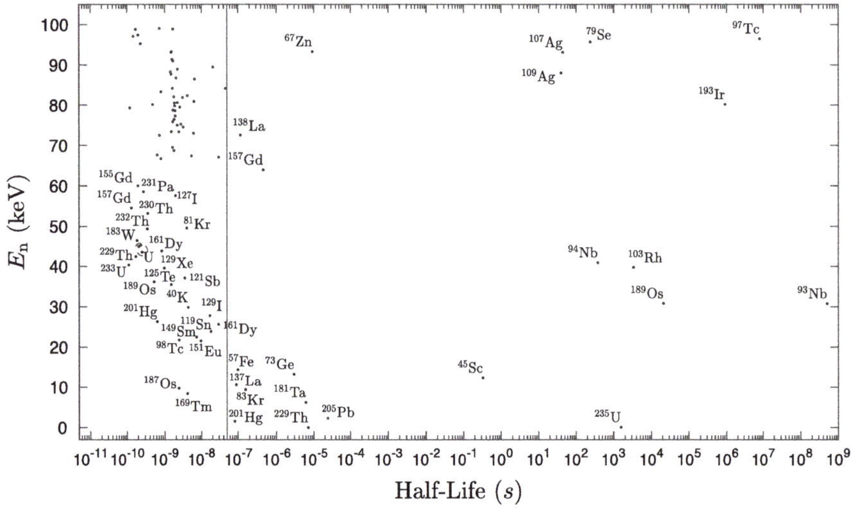

Fig. 5.5 Energy and half-life map of the isotopes meeting the search criteria. The labels of the isotopes in the upper-left corner have been omitted to improve readability. The gray vertical line is at 50 ns

Whether an isomeric decay is within experimental reach depends on the choice of the detector and the nature of the measured signal, i.e., converted electrons or γ-rays. In designing our experiment, we have chosen the Timepix3 [34], provided with a time resolution of 1.25 ns and a pixel energy threshold of $D_{\text{th}} = 3\,\text{keV}$. Therefore, with this device, it is possible to detect electrons, ions, and photons, provided that the energy deposited in each individual pixel exceeds 3 keV. However, if we consider the delayed nuclear decay shown in Fig. 5.3, a difference arises in relation to the pixel threshold. While photons emitted in the decay have an energy of $\hbar\omega = E_{\text{n}}$, the energy of converted electrons is reduced by the binding energy of the shell from which it originates, resulting in $E_{\text{ce}} = E_{\text{n}} - E_{\text{b}}$. This latter signal could potentially be lost for nuclear transitions close to the detector threshold ($E_{\text{n}} \sim D_{\text{th}}$). On the other hand, in terms of time resolution, one can expect the signal generated by plasma radiation to be broadened by the charge transport along the thickness of the silicon chip. This broadening could be of several hundreds of ns, strongly inhibiting the detection of short-lived excitations. Therefore, we can safely disregard from further consideration all the states having $T_{1/2} \leq 50\,\text{ns}$, resulting in a reduced set of 22 isotopes. Considering the excitations that can be realistically achieved with the characteristics of our laser, we need to further constrain the selection to transitions with $E_{\text{n}} \leq 20\,\text{keV}$.

Several of these highlighted isotopes, including ^{235}U, ^{57}Fe, ^{181}Ta, ^{201}Hg, have historical significance. The ease with which ^{181}Ta can be found commercially makes it a suitable test sample to prototype our experiment and obtain a comparison with the experiment in Ref. [24].

^{181}Ta has a $9/2^-$ excited level connected to the ground state by a 6.237 keV dipolar (E1) transition and a half-life $T_{1/2} = 6.05\,\mu\text{s}$: these properties are within reach for the Timepix3 detector. We have thus considered a solid cylinder with a 99.98% abundance of ^{181}Ta as a target for the optical laser-plasma experiment. The choice of ^{181}Ta also facilitates the selection of a control sample (i.e., one that will not produce a delayed nuclear decay) and makes it fall on tungsten (W). These two elements are consecutive on the periodic table and will have similar characteristics in terms of plasma formation and evolution. The composition of the W sample includes different isotopes according to their abundance. ^{183}W has an abundance of about 14% and has the lowest excitation level at 46.48 keV with a half-life $T_{1/2} = 0.185$ ns. The remaining three most abundant isotopes of W account for 85.57% and have their first excited state at about 100 keV with a half-life $T_{1/2} \sim 1$ ns. This is the reason why the signals generated by the two targets are expected to differ significantly in the time domain, even in the remote eventuality to induce a nuclear excitation in W.

5.2 Design of the Experimental Setup

The experimental apparatus, depicted in Fig. 5.6a, is situated within a square-shaped vacuum chamber with dimensions of 41.0 × 41.0 × 42.6 cm. The chamber is designed with a removable lid to facilitate access to its components.

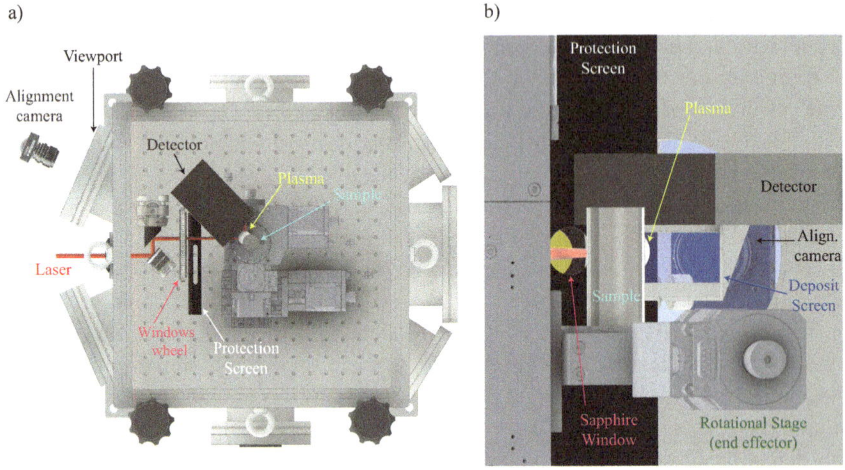

Fig. 5.6 3D CAD of the experimental setup designed at EPFL. In **a** the top view of the 3D CAD of the vacuum chamber. The experimental setup includes a four-axis mechanical stage, a Timepix3 detector, a plane and a parabolic mirror, a protection screen, and a motorized sapphire-windows wheel, all housed inside the vacuum chamber. The sample is placed on the rotational stage, and the detector is positioned to face the plasma plume generated by the laser, which mostly develops along the surface normal of the laser incident point. An alignment camera—placed outside the chamber—looks at the sample through the viewport flange to aid in alignment. In panel **b**, a closer view inside the chamber reveals the details of the protection and deposit screens. The protection screen has a through-hole covered by a sapphire window, which protects the parabolic mirror from deposition. The deposit screen is equipped with borosilicate glass, enabling inspection and alignment from the camera located outside the chamber

The sample is mounted on a four-axis motorized stage, which offers a rotational degree of freedom as its end effector. A laser source is directed into the chamber and subsequently redirected towards the sample through the use of a 90° off-axis parabolic mirror with a focal length of 10 cm. Upon illumination of the sample by a high-intensity fs-laser, a keV-hot plasma is generated. The plasma plume is known to expand predominantly along the surface normal of the laser incident point [35]. Therefore, the detector has been positioned in the direction of the plasma expansion. To mitigate the deposition of ablated nanoparticles on the optical components, a sapphire-windows motorized wheel, and a protection screen have been installed between the target and the parabolic mirror. Over time, the sapphire window can accumulate nanoparticles due to ablation, which may reduce the effective laser intensity reaching the sample. However, the motorized wheel enables the replacement of the sapphire window without requiring the chamber to be vented, thereby prolonging the overall lifespan of the experiment. The TimePix3 CCD is protected by a 25 or 50 μm layer of Kapton and placed downwards to prevent the deposition of nanoparticles on its surface. This layer has only a slight effect on X-ray absorption, while it can completely block electrons.

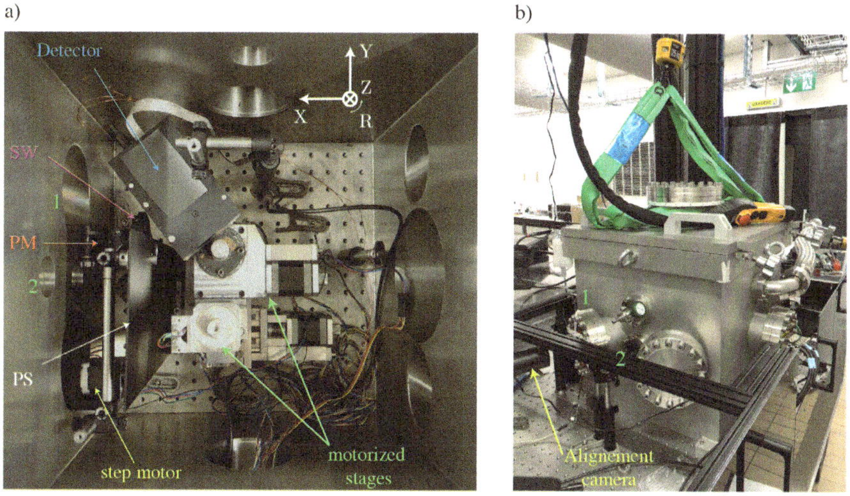

Fig. 5.7 Pictures of the experimental setup implemented at EPFL. In panel **a**, the inner part of the vacuum chamber is visible, with the sample placed on the four-axis motorized stage. The axes are labeled in white on the upper-right side. SW indicates the sapphire-windows wheel, PM the parabolic mirror, and PS the protection screen. In panel **b**, the vacuum chamber is seen placed on the optical table. The lid is attached to a crane for lifting. The numbers (1) and (2) refer to the viewport flanges used for the alignment (1) and for the laser insertion (2)

To address the potential issue of atomic recombinations happening at the chamber walls and to increase the detection angle of view for slower decaying nuclei (i.e., those that did not decay transiting from the target to the detector), a deposit screen has been incorporated into the design. This screen is placed in close proximity to the sample and directly on the detector itself, as illustrated in Fig. 5.6b. The central portion of this screen is composed of borosilicate glass, which facilitates visual inspection using the external alignment camera, as explained in the following section. The cylindrical sample is mounted on a four-axis motorized stage, which continuously moves to keep a fresh portion of the sample at the focal point. The rotational and Z-axis stages are responsible for refreshing the surface, while the X and Y axes maintain the cylinder surface under laser focus during the ablation process. Figure 5.7 shows pictures of the setup implemented at EPFL.

5.3 Alignment Procedure

Before the experiment can be performed, three crucial alignment steps must be completed. As it will become clearer in Sect. 5.4, these steps guarantee optimal plasma conditions—in terms of the highest temperature achievable—and stable counts throughout the measurements. These steps are:

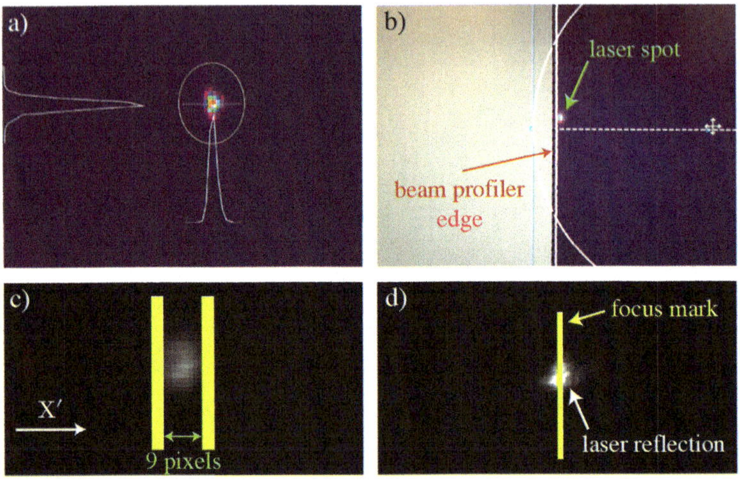

Fig. 5.8 Alignment procedures to be performed before the laser-plasma experiment. In **a**, the beam profiler is used to align the parabolic mirror to achieve the minimum possible spot size. **b** The Y-axis is driven to move the beam profiler so that the laser hits one of the lateral edges. **c** Laser reflection observed through the alignment camera during the self-alignment procedure. The spot has a width of 9 pixels ($\sim 31\,\mu m$), showing no further lateral movements. **d** The laser reflection is placed on the digital marker, which represents the position of the focus. This is achieved by driving the X and Y axes of the mechanical stage

1. Aligning the parabolic mirror to minimize monochromatic aberrations and produce the smallest possible focus spot;
2. Aligning the sample with the axis of rotation of the rotational stage;
3. Placing the sample surface at the laser focus and orienting it normally with respect to the direction of the detector.

For step 1, a beam profiler is positioned on the rotational stage (in place of the sample) to visualize the beam size, as depicted in Fig. 5.8a. Once the smallest focus has been obtained and the parabolic mirror aligned, the Y axis of the mechanical stage can be moved such that the laser spot hits the lateral edge of the beam profiler CCD (it is the beam profiler that moves and not the laser), as shown in Fig. 5.8b. At this point, the external alignment camera can be used to store the position of the focus through a digital marker. To effectively observe the edge of the CCD through the alignment camera, it is necessary to remove the protective neutral density (ND) filter from the beam profiler. When the target sample is positioned in place of the beam profiler, the laser impinging on its surface will produce a reflection that can be seen by the alignment camera, as shown in Fig. 5.8c, d.

At this point, the cylindrical sample can be accurately aligned along the axis of rotation of the mechanical stage through the utilization of a pointed metal rod while continuously rotating the cylinder (step 2). This rod is placed on a micro-stage and is gradually brought closer to the sample as the procedure goes on. Initially, the cylinder is misaligned, and the rotation of the stage will result in a contemporaneous shift in X and Y of the sample with respect to the axis of rotation. These deviations in the XY plane, combined with the curved surface of the cylinder, will change the position at which the laser impacts the target. As a consequence, the laser reflection observed through the alignment camera will begin to oscillate on the X′ axis (defined in Fig. 5.8c), with decreasing swing amplitudes as the cylinder self-aligns. The process is terminated when the spot remains still or the motion is only a small fraction of its dimension. For example, as shown in Fig. 5.8c, the reflection on the alignment camera has a size of 9 pixels, which corresponds to $\simeq$31 µm, and does not display any noticeable motion (boundaries are marked by the yellow vertical lines).

Finally, in step 3, the X and Y axes are adjusted to place the cylinder in the focus of the parabolic mirror. The position of the focus has been saved in step 1 through the digital marker, and now the laser reflection must be placed on it by moving the sample with the mechanical stages. In the scenario where the sample surface is flat and positioned perpendicular to the laser beam, a displacement of the sample along the Y-axis would not produce any change in the reflection observed by the alignment camera, as shown in Fig. 5.9. Therefore, the position along the X-axis of the mechanical stage at which the surface is in focus corresponds to a unique point on the X' axis of the alignment camera. This is why the digital marker is saved as a vertical line in Fig. 5.8d. In contrast, when using a cylindrical sample, both the X- and Y-axes need to be adjusted to position the cylinder at the laser focus. The impact point on the sample must be chosen in such a way that the detector is placed orthogonal to the surface. Furthermore, choosing an incident angle within the range of 25° to 45° decrease significantly—compared to normal incidence—the accumulation of nanoparticles on the sapphire window that protects the parabolic mirror.

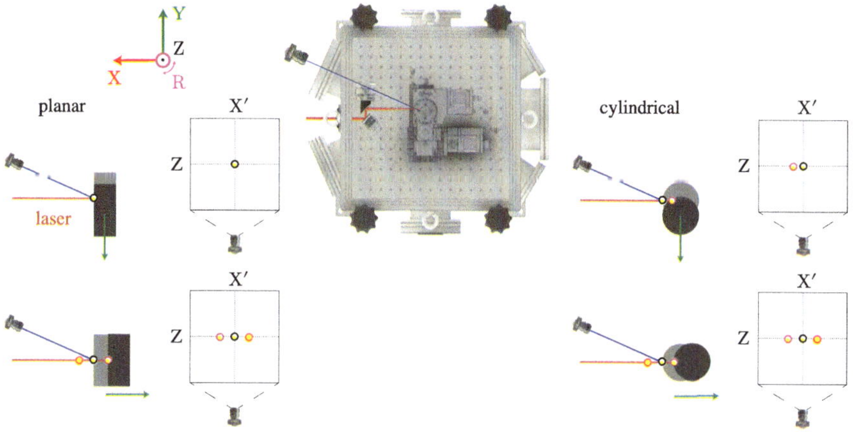

Fig. 5.9 Observing the movements of planar and cylindrical samples via the alignment camera. The blue line represents the optical axis of the camera, while the red line the laser beam. X, Y, Z, and R refer to the axes of the mechanical stage, while X′ is the horizontal axis of the alignment camera. The sample moves towards darker regions, as indicated by the green arrows

5.4 Plasma Hydrodynamic Model: Properties and Evolution of the Optical-Laser Generated Plasma

By examining the properties of the ultrafast laser used in these experiments [36], summarized in Table 5.1, it is possible to produce an initial estimate of the electron and photon distributions, along with the plasma temperature and other important plasma expansion properties. Even though the question may seem simple, the actual effects of short-pulse laser irradiation on a solid target are complex and cover a broad range of physical phenomena [3, 37–39]. When matter experiences interactions with high-intensity short-pulse lasers, regardless of the target type, a variety of physical processes take place, such as ionization, propagation, refraction, and plasma wave generation, as well as the consequent thermal and hydrodynamic expansion in the target material [3]. When the laser intensity exceeds $I \geq 10^{16}$Wcm^2, many of these effects depend on the laser irradiance[1] $I\lambda^2$. Experimental findings showed that collisional absorption alone could not have accounted for the significant rise in plasma temperature at irradiance conditions $I\lambda^2 \geq 10^{15}$ W cm^{-2} μm^2. In fact, at these irradiance levels, the rate of electron-ion collisions ($v_{ei} \sim I^{-3/4}$) diminishes significantly. This reduction is further amplified as the quiver velocity, which is the velocity with which the laser field drags the electrons, becomes comparable to the thermal velocity [3].

Therefore, other "collisionless" mechanisms needed to be identified in order to explain the high absorption observed in this regime. One such process is the resonance absorption in the case of obliquely incident laser fields [37, 40, 41]. Here, the component of the field that is perpendicular to the density profile has the potential to resonantly excite a plasma wave—oscillations of the free electrons—along its direction. This resonance appears as a singularity in the analytical expression of the electric field, which is located at the critical surface. The critical surface is defined as the surface where the electron density (n_e) equals the critical density ($n_c = {\epsilon_0 m_e \omega^2}/{e^2}$) such that the field cannot penetrate it. However, this resonance breaks for sharp-edge density profiles, as the amplitude of these oscillations (${v_{os}}/{\omega}$) exceeds the electron density gradient length (L). Another collisionless process was proposed by Brunel in

Table 5.1 E_{pulse} is the energy per pulse, f_{rep} is the repetition rate, λ is the laser wavelength, while D_{focal} is the FWHM of the laser spot at the focus of the parabolic mirror measured with a beam profiler. τ_{pulse} is the pulse duration at the chamber entrance. I is the laser peak intensity

E_{pulse}	1.25 mJ
f_{rep}	4 kHz
P	5 W
D_{focal}	10 μm
τ_{pulse}	31 fs
λ	0.78 μm
$I\lambda^2$	$\sim 3 \times 10^{16}$ W cm^{-2} μm^2

[1] Generally, the irradiance represents a radiant flux per unit surface; hence it is expressed in units of W m^{-2}. Here, we call irradiance the quantity $I\lambda^2$ in accordance with Ref. [3].

1987 [42]. In his model, the laser field directly pulls electrons away from the surface into the vacuum during the first half-cycle. During the second half-cycle, the electrons are driven back into the plasma with a velocity of approximately $v \sim v_{osc} = {}^{eE}/_{m_e\omega}$, where v_{osc} refers to the quiver velocity. This process was found to be more efficient than resonant absorption when $({}^{v_{os}}/_{\omega L} > 1)$, as it might be in the case of a femtosecond laser hitting a solid target. Specifically, for $\tau_{pulse} \sim 100$ fs, the target is heated at a rate faster than the hydrodynamic expansion timescale. Consequently, the plasma medium retains its initial sharp solid-state density profile throughout the duration of the laser pulse [37].

In this collisionless heating regime, the scaling of the electron temperature with respect to the laser irradiance can be inferred through models or electrostatic PIC simulations [43, 44]. In this context, we will examine the two scaling laws (SL) outlined in Ref. [3], which are the following:

$$T_e\ (\text{keV}) = \begin{cases} 3.6\ I_{16}\lambda_\mu^2 & \textbf{SL1} \\ 8\ \sqrt[3]{I_{16}\lambda_\mu^2} & \textbf{SL2} \end{cases} \tag{5.1}$$

These scaling laws, along with the hydrodynamic model presented in this section, have been applied in Refs. [9, 10] to determine the rates of nuclear excitation through photon absorption and NEEC in plasma generated by optical lasers. The intensity I_{16} in Eq. 5.1 is expressed in units of Wm^{-2}, while the wavelength λ_μ is expressed in µm. The plasma volume, V_p, generated by laser ablation can be described as a function of the focal radius, R_{focal}, and the pulse duration, τ_{pulse}, as follows:

$$V_p = \pi\ R_{focal}^2\ d_p = \pi\ R_{focal}^2\ c\ \tau_{pulse}\ . \tag{5.2}$$

Here, d_p represents the plasma thickness, which can be estimated as $c \cdot \tau_{pulse}$, with c being the speed of light in vacuum. The total number of electrons present in the neutral plasma, N_e, and the electron density, n_e, can be estimated as follows:

$$N_e = f\ \frac{E_{pulse}}{T_e} \quad \text{and} \quad n_e = \frac{N_e}{V_p}\ . \tag{5.3}$$

The number of free electrons in the plasma is determined by the absorption coefficient f, which—for the previously mentioned irradiance and steep density profiles—is approximately 0.11 [45]. When using the first scaling law (SL1), the number of free electrons is independent of the pulse energy E_{pulse}, as long as f remains relatively constant. By using the free-electron model, the chemical potential μ—appearing in the Fermi-Dirac distribution f_{FD}—can be obtained by imposing that the total number of electrons in the plasma is equal to n_e:

$$\int_0^\infty g_e(E_e)\ f_{FD}(E_e)\ dE_e\ = n_e\ , \tag{5.4}$$

and considering $E_e \gg \mu$, it results:

$$\mu = T_e \ln\left(\frac{2n_e}{\gamma\sqrt{\pi}(T_e)^{3/2}}\right). \tag{5.5}$$

The electron flux ϕ_e within the plasma is defined as follows:

$$\phi_e(E_e) = g_e(E_e)\, f_{FD}(E_e)\, v(E_e). \tag{5.6}$$

In this expression, g_e represents the electron density of states, f_{FD} represents the Fermi-Dirac distribution, and v represents the electron speed; all these variables are functions of the electron energy E_e. Under the assumption of thermodynamic equilibrium (TDE) between the photon and electron gases, the photon flux can be defined using a blackbody distribution as follows [9, 10]:

$$\phi_\gamma^{TDE}(E_\gamma, T_e) = \epsilon\, c\, g_\gamma(E_\gamma) f_{BE}(E_\gamma, T_e). \tag{5.7}$$

Here, g_γ represents the photonic density of states and f_{BE} represents the Bose-Einstein distribution [46]. Compared to the model presented in Refs. [9, 10], the emissivity ϵ has been introduced to impose the energy balance between the electron and photon gases. The energy contained in the photon gas is otherwise unrelated to the energy of the pulse and might exceed it. Both electron (ϕ_e) and photon (ϕ_γ) fluxes are expressed in units of $cm^{-2}\, eV^{-1}\, s^{-1}$. Another possibility is to consider the bremsstrahlung process as responsible for the origin of photons in the plasma. In this case, the emitted flux can be defined as [7]:

$$\phi_\gamma^B(E_\gamma, E_e, T_e, n_e) = t_i\left(\frac{d\sigma_B(E_\gamma, E_e)}{dE_\gamma}\right)\phi_e(E_e, T_e, n_e) \tag{5.8}$$

where $d\sigma_B(E_\gamma, E_e)/dE_\gamma$ is the differential cross-section of the bremsstrahlung emission of photons having energy E_γ by electrons of energy E_e and t_i is the target thickness expressed in atoms $\cdot$ m^{-2}. The values for the differential cross-section of Ta can be found in Ref. [47]. The target thickness $t_i = n_i R$ is given in terms of ion density n_i and plasma radius R. The two photon fluxes ϕ_γ^{TDE} and ϕ_γ^B are shown in Fig. A.5 calculated in the case outlined by SL1.

To determine the plasma lifetime, an estimate of the charge state distribution is necessary. This feature is also crucial for identifying the shells that can be targeted for electron capture. The radiative-collisional code FLYCHK [29] can be employed for this purpose once T_e and n_e are provided as input, assuming the plasma to be in its non-local thermodynamical equilibrium (non-LTE[2]) steady-state. Although steady-state conditions may not be reached for such short laser pulses (tens of fs),

[2] The non-LTE solutions are the most general descriptions of the thermodynamic conditions and population distribution of the plasma since the atomic processes are considered in detail through the solution of rate matrices [29].

this approach can still serve to obtain an initial estimate of several properties of the plasma. The results obtained from FLYCHK can be used to calculate the ion average charge state $\overline{Z}$, as follows:

$$\overline{Z} = \sum_q (q\ P_q)\,. \tag{5.9}$$

In this context, the variable q is used to denote the ion charge state, while P_q represents the probability of its presence in the plasma. The ion density (n_i) and the number of ions in the plasma (N_i) can be evaluated as follows:

$$N_\text{i} = n_\text{i}\ V_\text{p} \quad \text{and} \quad n_\text{i} = \frac{n_\text{e}}{\overline{Z}}\,. \tag{5.10}$$

To compute the hydrodynamic expansion of plasma, it is necessary to consider the energy conservation during collisions of cold ions and hot electrons in the expanding gases. The expansion rate is governed by the following non-linear differential equation, which is detailed in Ref. [4]:

$$\left(\frac{dR}{dt}\right)^2 = \frac{3\ T_{\text{e}0}\ R_0^2\ Z}{M_\text{i}}\left(1 - \frac{R_0^2}{R^2}\right), \tag{5.11}$$

under the assumption that the initial velocity of the plasma is equal to zero and that no electrons definitely escape from it. It is worth noting that the charge state Z in Eq. 5.11 hides an implicit dependence on the plasma radius R. By taking the smaller of the values between R_focal and d_p as the initial plasma radius R_0, a lower limit estimate can be obtained for the plasma lifetime (τ_plasma). The plasma lifetime is defined as the duration required by the plasma to double its initial radius and in our scenario, where $R_\text{focal} \leq d_\text{p}$, it can be expressed as:

$$\tau_\text{plasma} = R_\text{focal}\sqrt{\frac{m_\text{i}}{T_\text{e}\ \overline{Z}}}\,, \tag{5.12}$$

that is an approximate solution of Eq. 5.11. The electron temperature evolution during this adiabatic[3] expansion can be expressed as follows:

$$T_\text{e} = T_{\text{e}0}\left(\frac{R_0}{R}\right)^2. \tag{5.13}$$

In the following, I will discuss the implementation of this hydrodynamic model to the ^{181}Ta target and experimental setup that has been presented in this Chapter. However, it is worth mentioning that a more detailed description could be made by employing PIC simulations. The duration of the laser pulse, as reported in Table 5.1, is measured at the immediate output of the laser. Before interacting with the sample,

[3] Thus, this model does not account for the additional power radiated during the plasma expansion.

Table 5.2 Application of the plasma hydrodynamic model to our experimental conditions in case of a ^{181}Ta sample. Plasma properties obtained using the scaling laws (SL) outlined in Eq. 5.1 and the laser parameters reported in Table 5.1. $\phi_\gamma^{\mathrm{TDE,max}}$ refers to the maximum (peak) photon flux obtained using the blackbody distribution of Eq. 5.7, while $\phi_\gamma^{\mathrm{B,max}}$ represents the maximum photon flux originated by the bremsstrahlung process described in Eq. 5.8

	SL1	SL2
T_e	8.71 keV	10.74 keV
n_e	$1.05 \times 10^{20}\mathrm{cm}^{-3}$	$8.48 \times 10^{19}\mathrm{cm}^{-3}$
n_i	$1.86 \times 10^{18}\mathrm{cm}^{-3}$	$1.44 \times 10^{18}\mathrm{cm}^{-3}$
$\overline{Z}$	56.18	58.89
τ_{plasma}	9.79 ps	8.61 ps
ϕ_e^{max}	$2.72 \times 10^{25}\,\mathrm{cm}^{-2}\,\mathrm{s}^{-1}\,\mathrm{eV}^{-1}$	$1.98 \times 10^{25}\,\mathrm{cm}^{-2}\,\mathrm{s}^{-1}\,\mathrm{eV}^{-1}$
$\phi_\gamma^{\mathrm{TDE,max}}$	$5.40 \times 10^{25}\,\mathrm{cm}^{-2}\,\mathrm{s}^{-1}\,\mathrm{eV}^{-1}$	$3.60 \times 10^{25}\,\mathrm{cm}^{-2}\,\mathrm{s}^{-1}\,\mathrm{eV}^{-1}$
$\phi_\gamma^{\mathrm{B,max}}$	$6.75 \times 10^{18}\,\mathrm{cm}^{-2}\,\mathrm{s}^{-1}\,\mathrm{eV}^{-1}$	$4.31 \times 10^{18}\,\mathrm{cm}^{-2}\,\mathrm{s}^{-1}\,\mathrm{eV}^{-1}$

the laser beam is directed through four low-GDD ultrafast mirrors, a parabolic mirror, and two sapphire windows (yielding a total thickness of 5.5 mm). The pulse broadening is primarily attributed to the sapphire windows, as the GVD for sapphire is 49.9fs^2, resulting in a broadened pulse duration of $\tau_{\mathrm{pulse}} \simeq 40$ fs and a laser irradiance of $I\lambda^2 = 2.42 \times 10^{16}\,\mathrm{W\,cm^{-2}\,\mu m^2}$. The outcomes of this analysis, for both scaling laws, are presented in Table 5.2 and are found to be in agreement with each other. Regardless of the model chosen, it is crucial to make at least a consideration. In some circumstances, electron and photon fluxes might differ significantly by several orders of magnitude—generally to the advantage of photons [10]. Nuclear excitation processes, such as NEEC and direct photoexcitation, can occur as long as free electrons, vacancies, and photons are available. Since both electrons and photon fluxes persist for a duration assumed to be of the order of τ_{plasma}, the NEEC cross-section must be large enough to overcome this difference in the fluxes to become the dominant process. Otherwise, the NEEC excitation rate can be enhanced by supplying an external electron beam. In the case under study, the electron and photon fluxes are comparable in the case of TDE, while they differ significantly if the bremsstrahlung model is used. Which model will correctly represent the experimental conditions will depend on the dynamics and conditions of plasma formation and on the electron density (i.e., overdense or underdense plasma) [48, 49]. TDE conditions can be considered as the best-case scenario (upper limit), and consequently, so will the resulting photoexcitation rate. The results of the hydrodynamic model, when applied in accordance with scaling law SL1, are presented in Fig. 5.10. Specifically, Fig. 5.10a displays the ion charge state distribution probability (P_q) calculated using FLYCHK.

The temporal evolution of the plasma temperature and its radius is depicted in Fig. 5.10b, assuming a constant value of $Z = \overline{Z} = 56.18$ over the expansion. This approximation enables the calculation of the plasma lifetime as defined in Eq. 5.12. However, this picture is not entirely representative of reality, as illustrated in Fig. 5.10c, which shows the evolution of the average charge state throughout the

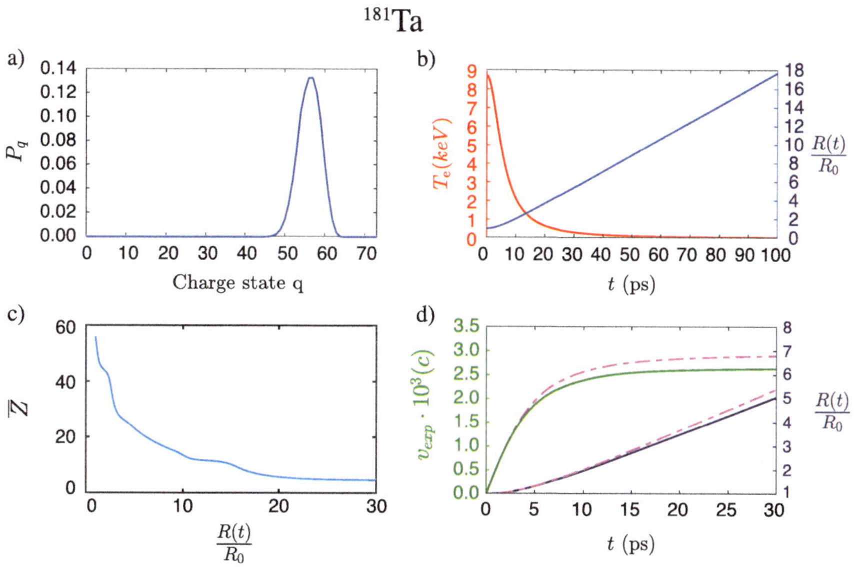

Fig. 5.10 **Plasma hydrodynamic model applied to the case of ^{181}Ta interacting with a femtosecond laser.** Panel **a** displays the ion charge state distribution calculated using the FLYCHK code. The temporal evolution of both the plasma temperature (in red) and radius (in blue) is displayed in panel **b**. The average charge state as a function of the plasma radius $R(t)$, and thus implicitly as a function of time, is presented in panel **c**. Finally, panel **d** displays the expansion velocity v_{exp} (in green), normalized to the speed of light, and the radius (in blue) obtained as solutions of Eq. 5.11, employing the average charge state $Z = \overline{Z}(R)$ presented in panel **c**. To facilitate comparison, the solutions obtained using a constant average charge state of Z=56.18 are displayed in dashed magenta lines

plasma expansion. This evolution has been computed using FLYCHK, with n_i and T_e provided as input for each value of the radius $R(t)$. In performing these calculations, it was assumed that the steady state for atomic processes is reached at each time instant over the expansion, which generally is slower and can last from tens to hundreds of ps [9, 10]. As anticipated, highly ionized atoms are predominantly present for $^{R(t)}/_{R_0} \leq 2$, with a significant drop in $\overline{Z}$ in the first hundreds of picoseconds. Figure 5.10d presents the solutions of Eq. 5.11 in terms of the radius and velocity of expansion calculated with two approaches: one using the time-dependent average charge state $Z = \overline{Z}(R)$ of Fig. 5.10c (solid lines), and the other obtained assuming $Z = \overline{Z} = 56.18$. Although differences are evident, the approximate solutions ($Z = \overline{Z} = 56.18$) presented in Fig. 5.10b are reasonably accurate.

The calculated average charge state of the ^{181}Ta ions is estimated to be around $\overline{Z} \sim 5$ using the SL1 model and $\overline{Z} \sim 11$ using SL2 at $^{R(t)}/_{R_0} = 30$, where the expansion velocity remains relatively constant, as shown in Fig. 5.10d. This indicates that in both models, the M shell is expected to be completely filled in hundreds of picoseconds or faster, whereas the N shell is only expected to be partially filled within the same time. Therefore, the models suggest that no increase (or only marginal) in the nuclear lifetime should be observed, as these two shells play a crucial role in determining the overall internal conversion coefficient $\alpha_{IC} = 70.5$ [50, 51]. This conclusion is in contrast with the findings presented in the study by Andreev et al. [24], where the measured time decay—depicted in Fig. 5.1—necessitates a ten-fold enhancement in the nuclear half-life to be attributed to the first excited level of ^{181}Ta. With this being said, a comprehensive PIC simulation will be useful to better describe the evolution and the relative charge state of the ions throughout the plasma expansion.

5.5 Nuclear Excitation Rates in Plasma Environment

By following the hydrodynamic model, it is possible to calculate the nuclear excitation rate resulting from the resonant absorption of a photon within the plasma with energy $E_\gamma = \hbar ck = E_n$ for the TDE and bremsstrahlung models, as follows:

$$\lambda_\gamma^{\mathrm{TDE}} = \int \sigma_\gamma(E_\gamma)\, \phi_\gamma^{\mathrm{TDE}}(E_\gamma, T_e, n_e)\, \mathrm{d}E_\gamma \,, \tag{5.14}$$

$$\lambda_\gamma^{\mathrm{B}} = \int\int \sigma_\gamma(E_\gamma)\, \phi_\gamma^{\mathrm{B}}(E_\gamma, E_e, T_e, n_e)\, \mathrm{d}E_\gamma\, \mathrm{d}E_e \,. \tag{5.15}$$

Here, σ_γ is the photoexcitation cross-section and it is expressed as [52]:

$$\sigma_\gamma(E_\gamma) = \frac{\pi^2\hbar}{k^2}\, W_\gamma^{\mathrm{E}\to\mathrm{G}}\, \frac{2J_{\mathrm{E}}+1}{2J_{\mathrm{G}}+1}\, L_r(E_\gamma - E_n) \,. \tag{5.16}$$

This equation links the photoexcitation cross-section σ_γ (which has a Lorentzian shape L_r centered at E_n) to the corresponding decay rate $W_\gamma^{\mathrm{E}\to\mathrm{G}}$ from the excited (E) to the ground (G) nuclear state, having angular momenta J_{E} and J_{G}, respectively. The decay rate $W_\gamma^{\mathrm{E}\to\mathrm{G}}$ is defined according to Eq. 2.16 for the electric and magnetic multipolar radiations. In first approximation, assuming a constant and homogeneous electron temperature T_e and density n_e over the plasma lifetime τ_{plasma}, the number of excited nuclei per laser pulse can be calculated as:

$$N_{\mathrm{exc}} = N_i\, \lambda_\gamma\, \tau_{\mathrm{plasma}} \,. \tag{5.17}$$

Subsequently, as illustrated in Fig. 5.3, the excited nucleus can decay via either γ-ray emission or internal conversion (IC). It is worth noting that the energy range of the converted electrons ejected from the M and N-shells—which contribute significantly to α_{IC}—resulting from the nuclear decay of the first isomeric state of ^{181}Ta are between 3.5 keV and 6.2 keV [33]. This range is above the detector threshold of $D_{\mathrm{th}} = 3$ keV. However, since we have protected the TimePix3 CCD with a Kapton

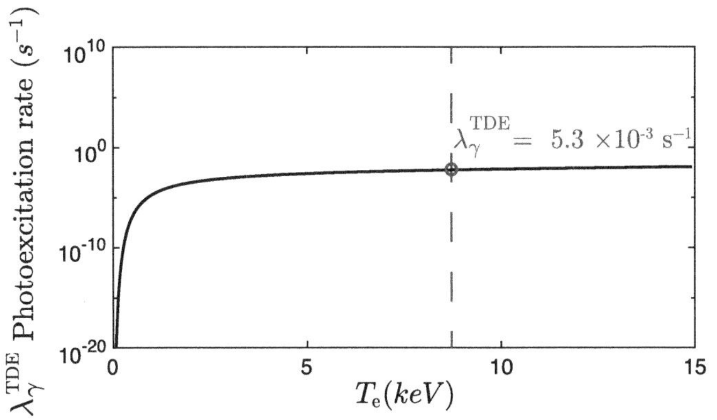

Fig. 5.11 Isomer photoexcitation rate as a function of the plasma temperature assuming TDE. Photoexcitation rate $\lambda_\gamma^{\rm TDE}$ of the 6.237 keV isomer in ^{181}Ta for plasma temperatures $T_{\rm e} \leq 15$ keV. The vertical dashed line corresponds to $T_{\rm e} = 8.71$ keV, which is the plasma temperature obtained with SL1. The photon flux of Eq. 5.7 has been used for this calculation

foil to prevent deposition, we can only be sensitive to γ-decays. Therefore, the number of radiative deexcitations can be estimated as:

$$N_{\rm deexc}^{\gamma} = \frac{1}{1+\alpha_{\rm IC}} N_{\rm exc} \,. \tag{5.18}$$

Figure 5.11 shows the photoexcitation rate $\lambda_\gamma^{\rm TDE}$ calculated for the E1 nuclear transition from the ground to the first isomeric state of ^{181}Ta—located at 6.237 keV—as a function of the plasma temperature $T_{\rm e}$, assuming the TDE photon distribution $\phi_\gamma^{\rm TDE}$ of Eq. 5.7. For comparison, the photoexcitation rate obtained using the bremsstrahlung photon flux distribution of Eq. 5.8—shown in Fig. A.5—is equal to $\lambda_\gamma^{\rm B} = 5.84 \times 10^{-11}\ {\rm s}^{-1}$ at $E_\gamma = E_{\rm n}$.

The outcomes of the SL1—when used in combination with the TDE assumption—indicate that each laser pulse results in approximately $N_{\rm exc} \sim 9 \times 10^{-5}$ excited nuclei and in $N_{\rm deexc}^{\gamma} \sim 1.27 \times 10^{-6}$ nuclei decaying by radiative emission. At a laser repetition rate of $f_{\rm rep} = 4$ kHz, the rate of formation of excited nuclei and the corresponding rate of deexcitation are $N_{\rm exc}^{\rm rep} \sim 0.36\ {\rm s}^{-1}$ and $N_{\rm deexc}^{\gamma,{\rm rep}} \sim 5.09 \times 10^{-3}\ {\rm s}^{-1}$, respectively. To provide a term of comparison, it is valuable to estimate the total number of photons emitted by the plasma surface throughout its lifetime as:

$$N_{\rm plasma}^{\gamma} = S_{\rm p}\, \tau_{\rm plasma} \int \phi_\gamma(E_\gamma, T_{\rm e}, n_{\rm e})\, dE_\gamma \,, \tag{5.19}$$

where $S_{\rm p}$ is the plasma radiating surface, which is assumed to be cylindrical with a radius $R_{\rm focal}$ and thickness of $d_{\rm p}$. By restricting the integration between $D_{\rm th} = 3\,{\rm keV} \leq E \leq 50\,{\rm keV}$, one obtains $N_{\rm plasma}^{\gamma} \sim 4 \times 10^{13}$. Therefore, the efficiency ratios are $N_{\rm exc}/N_{\rm plasma}^{\gamma} \sim 10^{-18}$ and $N_{\rm deexc}^{\gamma}/N_{\rm plasma}^{\gamma} \sim 10^{-20}$. A similar order of

Table 5.3 Number of excited nuclei, deexcitations, and photons in plasma, considering the TDE case. This table reports the number of excited nuclei (N_{exc}), the number of nuclei that deexcite through γ-decay ($N^{\gamma}_{\text{deexc}}$), and the total number of photons radiated by the plasma ($N^{\gamma}_{\text{plasma}}$) in the energy range $3\,\text{keV} \leq E \leq 50\,\text{keV}$. These values have been calculated per laser pulse and then scaled by the repetition rate f_{rep}

	Per laser pulse	$\cdot \times f_{\text{rep}}$ (s^{-1})
$N^{\text{TDE}}_{\text{exc}}$	9.10×10^{-5}	0.36
$N^{\gamma,\text{TDE}}_{\text{deexc}}$	1.27×10^{-6}	5.09×10^{-3}
$N^{\gamma,\text{TDE}}_{\text{plasma}}$	4.34×10^{13}	1.73×10^{17}

Table 5.4 Number of excited nuclei, deexcitations, and photons in plasma, considering bremsstrahlung as the origin of the photons gas. This table reports the number of excited nuclei (N_{exc}), the number of nuclei that deexcite through γ-decay ($N^{\gamma}_{\text{deexc}}$), and the total number of photons radiated by the plasma ($N^{\gamma}_{\text{plasma}}$) in the energy range $3\,\text{keV} \leq E \leq 50\,\text{keV}$. These values have been calculated per laser pulse and then scaled by the repetition rate f_{rep}

	Per laser pulse	$\cdot \times f_{\text{rep}}$ (s^{-1})
$N^{\text{B}}_{\text{exc}}$	1×10^{-12}	4×10^{-9}
$N^{\gamma,B}_{\text{deexc}}$	1.4×10^{-14}	5.6×10^{-11}
$N^{\gamma,B}_{\text{plasma}}$	3.98×10^{5}	1.59×10^{9}

magnitude for $N^{\gamma}_{\text{plasma}}$ could have been obtained by starting with the Stefan-Boltzmann blackbody emissive power $P = \epsilon\sigma T_{\text{e}}^4$ [53] radiated by the plasma surface S_{p} over a time τ_{plasma} and then normalizing for the weighted average energy of the emitted photons. For simplicity, the values discussed here have been reported in Table 5.3 for the TDE case and in Table 5.4 for the bremsstrahlung model.

Another mechanism that might enable the excitation of the nuclei in plasma is the process of nuclear excitation by inelastic electron scattering (NEIES) [13, 14, 54]. Here, the nucleus is excited via a free-free electronic transition involving the free electrons present in the plasma. The excitation rate can be defined as:

$$\lambda_{\text{NEIES}} = \int \sigma_{\text{NEIES}}(E_{\text{e}})\, \phi_{\text{e}}(E_{\text{e}}, T_{\text{e}}, n_{\text{e}})\, \mathrm{d}E_{\text{e}}, \tag{5.20}$$

where σ_{NEIES} is the NEIES cross-section taken from Ref. [13] and ϕ_{e} is the electron flux. Table 5.5 reports the excitation rate considering the plasma condition outlined by SL1 for the NEIES process.

When comparing the values across the three tables, it becomes evident that the photoexcitation process under the TDE conditions yields the highest excitation rate. Therefore, $N^{\gamma,\text{TDE}}_{\text{deexc}}$ can subsequently serve as a benchmark (upper limit) to assess whether the radiative decay of the nuclei can be observed experimentally. The effi-

Table 5.5 Number of excited nuclei and radiative deexcitations, considering the nuclear excitation by inelastic electron scattering (NEIES) as excitation process. This table reports the number of excited nuclei (N_{exc}), the number of nuclei that deexcite through γ-decay ($N^{\gamma}_{\text{deexc}}$). These values have been calculated per laser pulse and then scaled by the repetition rate f_{rep}. For these calculations, it has been used the NEIES cross-section present in Ref. [13]

	Per laser pulse	$\cdot \times f_{\text{rep}}$ (s^{-1})
$N^{\text{NEIES}}_{\text{exc}}$	7.70×10^{-6}	3.10×10^{-2}
$N^{\gamma,NEIES}_{\text{deexc}}$	1.08×10^{-7}	4.32×10^{-4}

ciency ratio for NEIES should be referenced in relation to electrons, not photons. If we were to factor in photons, the ratio would depend on the specific model employed for the photon gas. In this latter case, for bremsstrahlung one obtains $N^{\gamma,\text{NEIES}}_{\text{deexc}}/N^{\gamma,\text{B}}_{\text{plasma}} \sim 10^{-13}$, while $N^{\gamma,\text{NEIES}}_{\text{deexc}}/N^{\gamma,\text{TDE}}_{\text{plasma}} \sim 10^{-21}$ for TDE conditions. This indicates that, in the bremsstrahlung model, NEIES plays a more dominant role than direct photoexcitation.

5.6 Experimental Prototyping

In the early phases of the experimental setup characterization, measurements were conducted using the AdvaPix TPX3 [55] in the configuration depicted in Fig. 5.7. The spectra acquired through the "spectral imaging" function over a duration of 500 s with a laser power of 5 W and repetition rate of 4 kHz is shown Fig. 5.12a.

The acquired spectra exhibit an excellent match with the black-body distribution B_λ for energies greater than 3 keV. This energy corresponds to the pixel threshold D_{th} (denoted by a green vertical line), below which the detector disregards the event. The fit reveals a temperature of the distribution of $T_\gamma = 0.89$ keV. The measured temperature T_γ is much lower than the electron temperature T_e estimated based on the scaling laws presented in Table 5.2. This can be attributed mainly to three factors. (i) Firstly, the measured radiation results from averaging over the plasma lifetime. (ii) Additionally, the photon gas might not be in thermal equilibrium with the electron gas, in opposition to the hypothesis of the TDE case. (iii) Finally, the interplay between the pixel threshold and charge sharing, as discussed in the literature by Jakubek et al. [56], can result in the loss of events, leading to distortions in the measured spectrum and in a lower measured T_γ. Figure 5.12b displays a histogram of the recorded events as a function of their time of arrival (ToA) relative to the laser trigger. Despite the plasma having an emission lifetime on the order of hundreds of picoseconds, the trace left on the detector is approximately 40 ns in FWHM and has a total temporal spread of 236 ns. The response observed in these measurements, which in the first approximation is attributed solely to plasma radiation, shows that after a few hundreds of ns the initial plasma afterglow does not produce any appreciable effect. This is particularly important since the ToA is a critical observable that enables

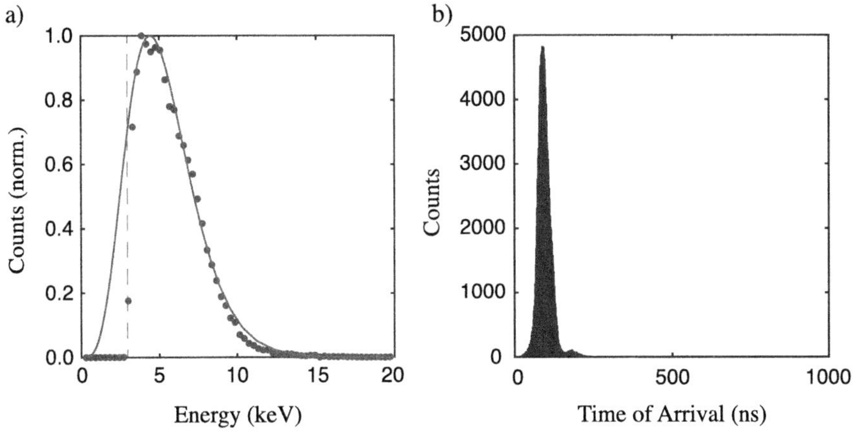

Fig. 5.12 Experimentally measured plasma radiation as a function of energy and time. Panel (**a**) shows the spectra acquired with a laser power of 5 W (dots) and the relative Planck's B_λ distribution fit with $T_\gamma = 0.89$ keV (solid curve). The frequency parametrization B_ν yields a higher temperature $T_\gamma = 1.6$ keV, but is less accurate in fitting the measured spectrum. A dashed vertical line has been included to indicate the pixel threshold D_{th} of 3 keV. In panel (**b**), the histogram for each recorded event is shown with respect to the time of arrival. This histogram has been extracted from a very narrow integration window

differentiation between plasma radiation and nuclear decay. These measurements also confirm—as initially expected—that despite the time resolution of the detector being of about 1.25 ns, only nuclear decays that have $T_{1/2} \geq 100$ ns could be easily detected.

With each laser pulse, a portion of the sample material is ablated and mainly ends its journey forming a thin film on the deposit screen, shown in Fig. 5.13a, b as it looks before and after the experiment. The ablation process will leave grooves on the sample surface, visible in Fig. 5.13c. The film deposition, which is effectively a pulsed laser deposition (PLD) [57], will also affect the sapphire windows placed on the motorized wheel behind the protection screen, shown in Fig. 5.13d. The motorized wheel can be operated whenever the detected counts decrease significantly: this is an indication that a new sapphire window should be placed as protection for the optics. Figure 5.13e shows instead the region of photoemission on the sample (i.e., where the laser impact on it), obtained at low laser intensity (hundreds of mW). The rotational and Z-axis stages are in charge of doing a surface refresh continuously. These expedients make it possible to keep the detected events almost constant for several hours. Additionally, if required, the movement of the XY stages can be employed to maintain the outermost portion of the surface within the laser focus once the entire length of the sample has been scanned.

To ensure accurate positioning and spacing of the laser pulses along the Z-axis and XY plane, as well as to validate some of the assumptions made in the hydrodynamic model, it was necessary to characterize the ablated surface. To accomplish this, profile

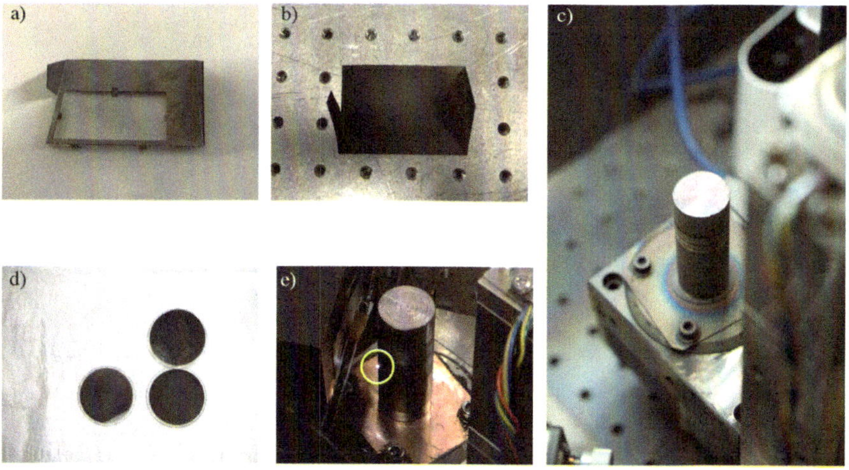

Fig. 5.13 Pictures showing film deposition on some of the components of the experimental setup. Panels **a** and **b** show the deposit screen placed on the detector before and after running the experiment. From panel **b**, it is evident that a tantalum film was deposited. Panel **c** shows the tantalum cylindrical sample on the rotational stage. Grooves left from laser ablation are clearly visible in the upper part. Panel **d** shows the sapphire windows with tantalum film deposition. In panel **e**, the region where the laser impacts the sample, resulting in photoemission, is indicated by a yellow circle

measurements were performed using a 3D optical profilometer on the tungsten (W) sample.[4] The profiles obtained are shown in Fig. 5.14, which displays the outcome for three fluences: 1 W (red), 3 W (blue) and 7 W (purple) at a repetition frequency of 4 kHz. The depth and width of the grooves formed by the laser beam were determined by applying a median filter to the profiles following the alignment of their minima at each y-position. The median and 95% percentile of these profiles, across all 1200 y-positions, are presented in Fig. 5.14d. The FWHM of the grooves was found to be 9.81 μm for a fluence of 1 W, 9.68 μm at 3 W and 15.13 μm for a fluence of 7 W. The traces left for 1 W and 3 W have lateral dimensions that are comparable with FWHM of the laser spot at the focus of the parabolic mirror ($2R_{\mathrm{focal}} \sim 10\,\mu\mathrm{m}$), as shown in Fig. 5.15a. However, for the highest possible fluence at 4 kHz, which is 7 W, the width of the groove is noticeably larger. This can be attributed to the monochromatic aberrations of the parabolic mirror, particularly to coma [58].

According to the hydrodynamic model presented in Sect. 5.4, the width of the grooves provides insight into the maximum temperature reached by the plasma during its formation, while the depth acts on the determination of the plasma volume (V_p) and in turn on the electron density (in SL1 the n_e is independent of R_{focal}) as seen in Eq. 5.3. Depths extracted from the 3D profiles are shown in Fig. 5.15b. They result to be 26.41 μm for 1 W, 35.67 μm for 3 W and 44.67 μm for 7 W. The sur-

[4] The Ta sample had already grooves along its entire length, and it was not possible to relate the size of the grooves with the laser power.

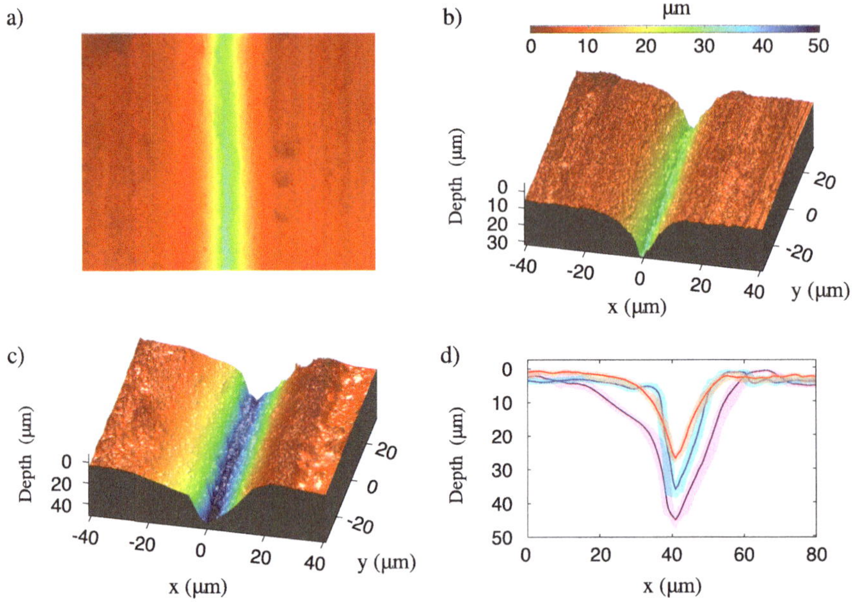

Fig. 5.14 Surfaces of the W cylindrical sample acquired through 3D optical profilometer. A 2D **a** and 3D **b** maps of an 80 × 60 μm section of the W (tungsten) sample. This section of the sample was irradiated with a laser power of 1 W. Panel **c** displays the 3D profile of the groove left by the laser at a power of 7 W. The colormap is shared among panels **a**, **b** and **c**. 3D profiles of the tungsten surface, such as those shown in **b** and **c**, are used to extract the median (solid curve) and 95% percentile (shaded area) plotted in panel **d** for laser powers of 1 W (red), 3 W (blue), and 7 W (purple)

face motion of the sample is estimated to be no greater than 0.6 μm per laser pulse at 4 kHz. Due to this limitation, only a small fraction of the sample is refreshed between two consecutive pulses. As a result, the groove depth cannot be attributed to a single laser pulse. Similar considerations may also apply to the full width at half maximum (FWHM). Considering the depth of focus of the beam (∼ 80 μm) and the fact that it takes approximately 20 pulses to move the surface by one spot dimension (D_{focal}), it can be concluded that the grooves are likely the result of multiple laser pulse hits. For comparison, the value of the plasma thickness $d_p(= c \times \tau_{\text{pulse}})$, used in the Eq. 5.2 of the hydrodynamic model, has been added in Fig. 5.15b. The definition of d_p is independent of the laser power and ranges between one-fourth to one-half of the measured depth of the groove across the different fluences. The considerations outlined in this Section, along with the other assumptions discussed in Sect. 5.4, suggest that while R_{focal} and d_p are consistent with the observed grooves on the sample, the hydrodynamic model may overestimate the actual experimental conditions that are present in the laboratory.

Preliminary time series (15 s each) acquired for the tantalum and tungsten samples are shown in Figs. 5.16 and 5.17, respectively. One can immediately distinguish a

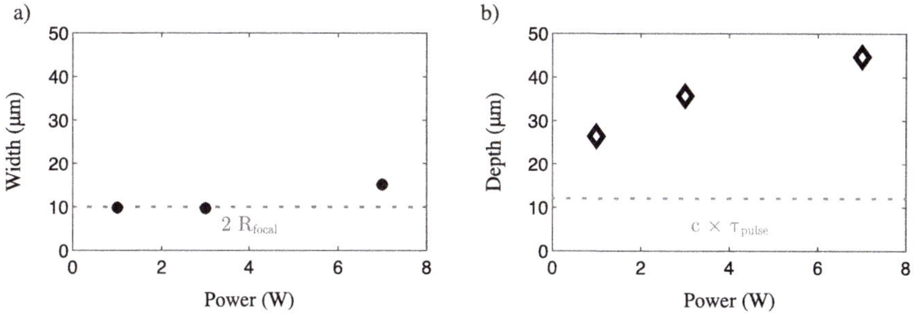

Fig. 5.15 Widths and depths of the grooves left by laser ablation on the W sample. Panel **a** displays the FWHM of the three profiles reported in Fig. 5.14d, plotted as a function of the laser power. For comparison, the FWHM of the laser spot measured at the focal point (D_{focal}) is added as a horizontal dashed line. The groove depths are shown in **b**, along with the value of d_{p} represented as a horizontal dashed line

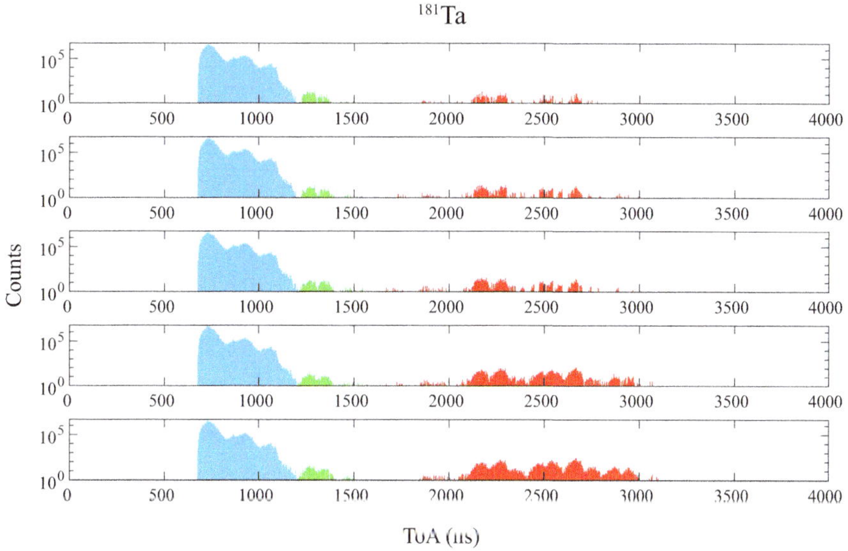

Fig. 5.16 Optical-laser generated plasma: time series acquisitions with the Tantalum sample

sharp peak with multiple bumps due to the plasma radiation (in blue), which could be overimposed to the time response of the silicon detector. A delayed response (shown in red) is also visible, likely originating from the plasma plume reaching the deposit screen positioned beneath the detector or hitting the chamber walls. A third feature, shown in green, accompanying the initial plasma radiation may also be present. It is possible to see that despite elements of high repeatability among different time series acquired at nominally equal laser conditions, there is a visible variation in the number of total counts measured in the region $2\,\mu\mathrm{s} \leq \mathrm{ToA} \leq 3\,\mu\mathrm{s}$ of Fig. 5.16.

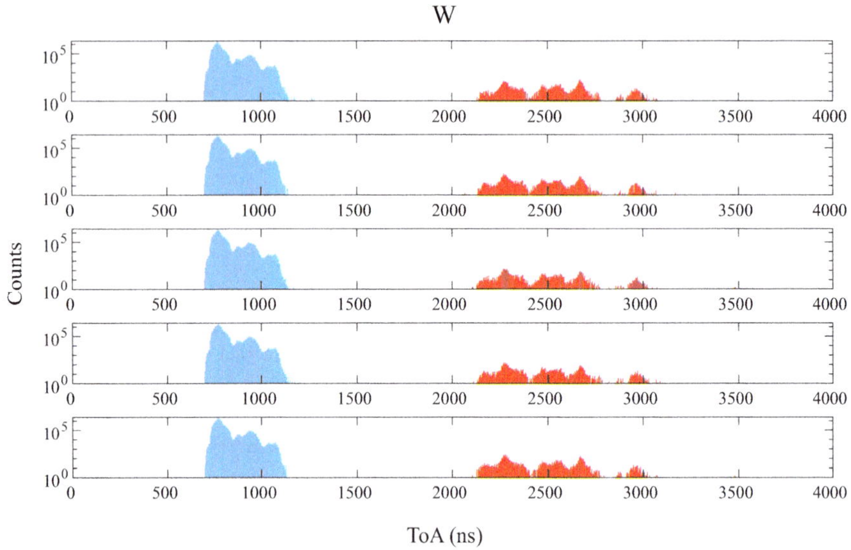

Fig. 5.17 Optical-laser generated plasma: time series acquisitions with the Tungsten sample

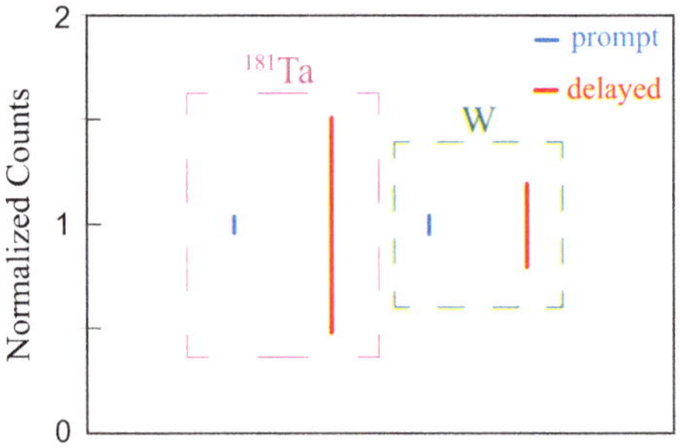

Fig. 5.18 Count dispersion among different time series. The 68% confidence interval is obtained by integrating the counts of Figs. 5.16 and 5.17 over two different time intervals: $0.5\,\mu s \leq$ ToA $\leq 1.7\,\mu s$ for prompt radiation (blue) and $1.7\,\mu s <$ ToA $\leq 3.5\,\mu s$ for the delayed events (red). The prompt and delayed integrated counts for each time series are then normalized to their respective average counts calculated across all the series. The bars represent the dispersion range of the normalized counts with a 68% confidence level for the Tantalum (Ta, in a magenta box) and Tungsten (W, green box) samples

Suppose we integrate the total counts present in the intervals $0.5\,\mu s \leq$ ToA $\leq 1.7\,\mu s$ (prompt, integrated counts $\simeq 10^7$) and $1.7\,\mu s <$ ToA $\leq 3.5\,\mu s$ (delayed, integrated counts $\simeq 10^3 - 10^4$) among different time series, and then we normalize to their average, we obtain the 68% confidence interval shown in Fig. 5.18. This demonstrates that even though the X-ray yield of the prompt plasma radiation is comparable among the five series shown in Figs. 5.16 and 5.17, the detected counts in the delayed region can exhibit a broader variation in relative terms.

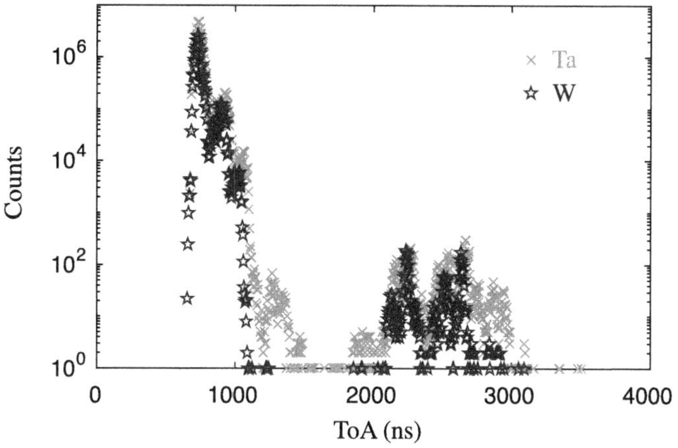

Fig. 5.19 Comparing the time-dependent signals measured with the Tantalum and Tungsten samples. The last time series in Figs. 5.16 and 5.17 have been superimposed and aligned to the peak of the prompt radiation to facilitate comparison. The two signals have not been normalized to each other, highlighting the high repeatability of the experimental conditions

These findings suggest that if the signal emitted by the ions interacting with the metallic walls of the chamber extends over the time interval when nuclear decay is expected, as it happens in Ref. [24], the excess of counts cannot be reliably used to discriminate the nuclear deexcitation. Instead, the time response and the energy spectrum must be considered as the distinctive signature of the process. In Fig. 5.19 the two last time series of Figs. 5.16 and 5.17 are superimposed and aligned in time to the peak of the prompt region to facilitate comparison. This confirms that, while being measured in different moments, the plasma produced with Ta and W are quite similar in terms of counts and overall dynamics of the plasma (for example, the ToA of the delayed X-rays).

If the delayed portion of the signal—depicted in red—is interpreted as due to ions hitting and interacting with the deposit screen,[5] it would suggest that the plasma expansion occurs at much lower velocities than those predicted by the hydrodynamic model presented in Sect. 5.4. Solution of Eq. 5.11, shown in Fig. 5.10, predicts an ion expansion velocity of $v_{\text{exp}} = 2.63 \times 10^{-3}c$ at ${}^{R(t)}/_{R_0} = 30$. This corresponds to a travel time of approximately 70 ns for the plasma to cover the distance between the sample and the detector, which is between 4 cm and 6 cm, and approximately 250 ns to reach the chamber walls. On the other hand, the model suggests that an eventual signal resulting from the ions traveling at v_{exp} and reaching the detector would fall well within the time window of the prompt radiation. To better comprehend the structure of the delayed events, the time interval between the peak of the prompt radiation and the subsequent peaks in the delayed events was calculated. The time

[5] Further discussion on this topic can be found in the following section.

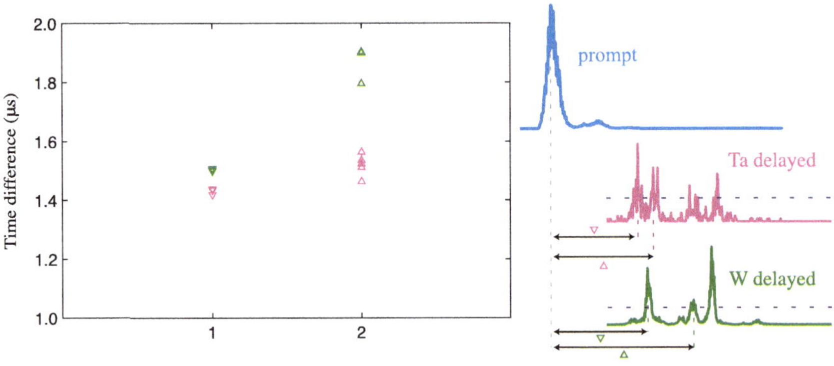

Fig. 5.20 Time difference between the prompt peak and the following peaks in the delayed events. The time intervals between the prompt radiation and the first peak of the delayed events are indicated with ▽ and plotted along the label '1'. Those relative to the second peak are indicated with △ and plotted along the label '2'. The horizontal gray dashed lines in the figure indicate that the first and second peaks for the delayed events were defined using a threshold of 35% of the maximum value measured in the corresponding time interval for each time series. The time intervals for Ta have been plotted in magenta, while those for W are in green. As one would expect, the interval with respect to the first peak is a much more reliable measure, showing high repeatability between Ta and W. Prompt and delayed signals, shown on the right side, are not to scale

distance between the prompt and the first (1) and second (2) delayed peaks are shown in Fig. 5.20.

The time intervals measured are on the order of $\sim 1.5\,\mu\text{s}$ both for Ta and W, which is roughly twenty times larger than the time predicted by the hydrodynamic model to reach the deposit screen and roughly six times larger the time predicted to reach the chamber walls. If these counts were associated with the ions of the primary hot plume, it would indicate a $T_e \ll 500\,\text{eV}$, which seems not compatible with the measured spectra. Therefore, these delayed events are presumably associated with the ions interacting with the chamber walls. Another possible explanation could be the presence of hot and cold ion distributions within the plasma. This raises the question of whether the green peak observed in Fig. 5.16 and other bumps in the blue prompt radiation could be attributed to the portion of plasma traveling at $v_{\text{exp}} \sim 3 \times 10^{-3}c$—and thus reaching the screen within hundreds of ns, while the red delayed events occurring after $\sim 1.5\,\mu\text{s}$ associated with the cold bulk part of the plume. A 2D energy-time histogram (discussed in the next section) and PIC simulations could help to shed light on this question. However, to uniquely identify these contributions, further analyses in various geometries are required.[6]

[6] Upon completing the original thesis, I conducted a detailed analysis of the CCD pixel locations responsible for these delayed counts. This showed that their occurrence is due to the detector's electronics being unable to keep pace with the readout demands in the presence of high-intensity radiation. This artificial afterglow disappears when lower laser intensities are used. An update on these experimental measurements can be found in Ref. [61].

5.7 Experimental Results

After following all the procedures discussed in this chapter, the experiment—aimed at the observation of the isomer decay—can proceed by acquiring data for several hours. At the time I started these measurements, the laser had a maximum output power of 5 W at a repetition rate of 4 kHz. As indicated in Table 5.1, the pulse duration was measured to be 31 fs at the laser output, and the FWHM of the focal spot on the sample was 10 µm, which is very close to the diffraction limit. A digital pixel mask was prepared to immediately remove false events due to defective pixels. In the case of uniform illumination, as should be the case in our experiment, a flat histogram of occurrences relative to the pixel number is expected. Thus, any events originating from pixels with substantially higher statistics than their surroundings were disregarded. Histograms illustrating the distribution of the detected events as a function of the pixel number before and after pixels removal are shown in Fig. A.2. AdvaPix TPX3 does not provide the capability to start a new measurement every 250 µs, which is the period between two consecutive laser pulses. Instead, it offers the option to refresh the 14-bit ToA counter through an external trigger. In our case, this signal has been driven by the laser trigger. Over the course of $\sim$ 10 h, a total of approximately 17 billion counts were collected, as depicted in Fig. 5.21. During this time, the sapphire window protecting the optics was replaced three times with the motorized wheel as soon as the rate was observed to decrease. A substantial fraction of these 17 billion counts[7] originated from pixels considered malfunctioning after plasma radiation, and therefore possessing higher statistics than their surrounding. Some of these events, despite originating from pixels with regular activation statistics (see Fig. A.2), were characterized by values of ToA preceding the arrival of the laser pulse on the sample. These events consistently showed a ToA value ~ 0 (in the detector units), even when the laser trigger was shifted. This could indicate a malfunction in the detector electronics, probably caused by the high photon flux. Additionally, the built-in software of the AdvaPix TPX3 does not handle energy and time information directly, therefore to obtain an energy-time map it is necessary to apply an energy calibration function to each event. This energy calibration function depends on the particular pixel from which the event originated.

Figure 5.22 shows the reconstructed 2D energy-time histogram obtained by considering a time window of 10 µs—starting few hundreds of ns before plasma onset—and an energy window of 50 keV. A dashed red line indicates the energy at which the nuclear decay is expected. The total number of 6.2×10^9 events[8] contained in this window seems compatible with the bremsstrahlung photon flux described in Eq. 5.8. Considering the rate of radiated photons $N_{\text{plasma}}^{\gamma,\text{B}} = 1.59 \times 10^9\ \text{s}^{-1}$ within the energy range $3\,\text{keV} \leq E_\gamma \leq 50\,\text{keV}$ (as determined in Table 5.4), along with a solid angle

---

[7] This is the total number of events detected before the pixels removal shown in Fig. A.2 was applied.

[8] This is the total number of unclustered events; therefore, an N-pixel trace generated by a single photon will result in N events. According to the AdvaPIX TPX3 built-in clustering software, our preliminary data (the code can handle only a small subset compared to Fig. 5.22) were strongly dominated by 1-pixel traces, thus we assume in the first approximation that $N_{\text{events}} \sim N_{1\text{pix}}$.

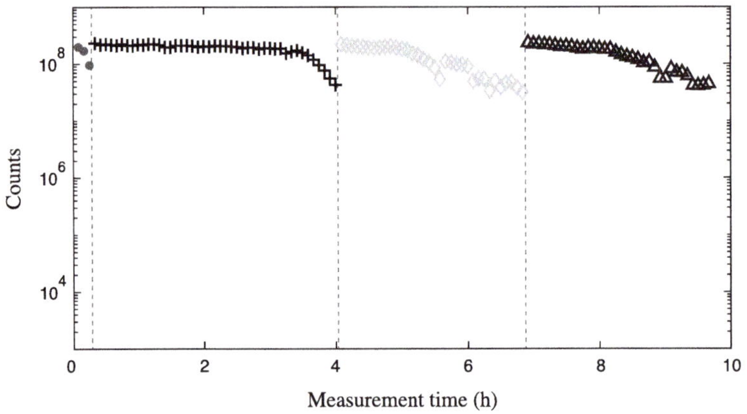

Fig. 5.21 Optical-laser generated plasma experiment: total number of events over time. Each marker represents the total number of events detected within a time window of 300 s. Different markers represent runs obtained after replacing the sapphire window with a clean one, and a dashed vertical line indicates the replacement. The total number of detected events from the raw data amounts to $\sim 1.7 \times 10^{10}$, collected over a total of 116 time series. Compared to the others, the first sapphire window had to be replaced quite early. This was probably due to an intensified accumulation of Ta nanoparticles on the sapphire window, potentially influenced by the surface properties of the sample or laser pointing

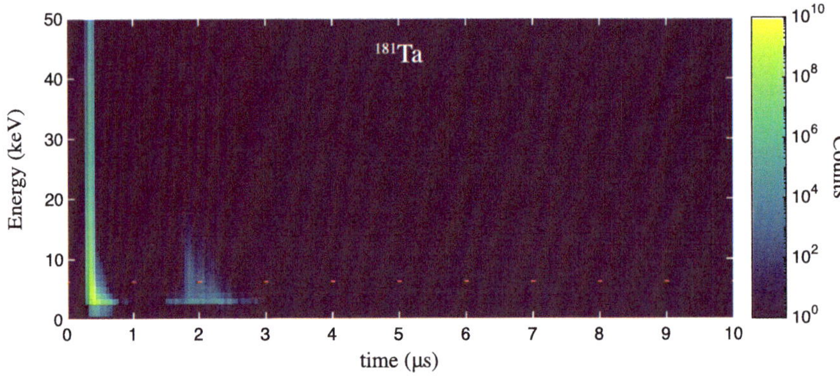

Fig. 5.22 Optical-laser generated plasma experiment: 2D energy-time map. The time bin has been chosen to be 50 ns, while the energy bin is of 1 kev. The total number of events in this image is approximately equal to 6.2e9. The red dashed line indicates the energy position at which it would have been expected to see the nuclear decay of the first isomeric state of ^{181}Ta. Since the color axis starts from 10^0, energy-time bins with a single count are indistinguishable from those with zero counts. This choice has been made to facilitate readability. However, in Fig. A.3, the same histogram is replicated using a different colormap and a lower limit for the color axis

of $\Omega \sim 0.04$ sr compared to 4π, we can estimate a total photon count of $\sim 2 \times 10^{11}$ for an integration time of 10 h. It is important to note that the photon rate calculated in Table 5.4 assumes constant plasma radiation over its lifetime τ_{plasma}. In reality, T_e and, consequently, ϕ_γ will significantly decrease over this timescale. This, along with the other hypothesis discussed in Sect. 5.4, may lead to overestimation. Detector efficiency and absorption of the 25 µm Kapton foil (used to protect the CCD) have only a minor impact on correcting the total counts, which becomes 7.4×10^9 after efficiency adjustments and 9.3×10^9 accounting also for the transmittance of Kapton. Although the bremsstrahlung model roughly reproduces the number of events detected, it does not reproduce the measured photon distribution, as can be seen in Fig. A.6. The energy spectrum is more accurately described by a Planck distribution B_λ with a specific emissivity ϵ, as can be deduced from Fig. 5.12, and later from Fig. 5.24.

A broad-energy signal associated with the prompt plasma radiation is observed in Fig. 5.22 at the beginning of the window. After approximately 1.5 µs, delayed events are detected, less intense and less extended in energy than the initial radiation. It appears that no signal related to a decay consistent with $T_{1/2} \sim 6$ µs and $E \sim 6$ keV, which might be associated with the ^{181}Ta isomer, is observed.

There may be several reasons that could explain the absence of a signal originating from nuclear deexcitation. The first potential explanation arises from the theoretical predictions discussed in Sect. 5.4. If the hydrodynamic model presented accurately describes the conditions of our experiment, the calculated efficiency ratios for direct photoexcitation according to Tables 5.3 and 5.4 are expected to be $N^\gamma_{\text{deexc}}/N^\gamma_{\text{plasma}} \sim 10^{-20}$. These values might be additionally corrected by accounting for the difference in the solid angles with which we detect plasma radiation and nuclear deexcitation, which are in a ratio of $\Omega^{\text{deexc}}_\gamma/\Omega^{\text{plasma}}_\gamma \sim 10$. This implies that in order to observe a single γ-decay due to direct photoabsorption, we would need to integrate[9] for $10^9\text{h} - 10^{10}\text{h} = 1\text{My}$. When considering NEIES, this period reduces drastically to a range of 100 and 1000 h, still for just one nuclear radiative deexcitation. However, this is still notably longer than our measurement window. Therefore, a nonobservation of the nuclear radiative deexcitation in our experiment is consistent with the theoretical predictions presented in Sect. 5.5.

These findings also raise the question of whether a signal related to the decay of the ^{181}Ta isomer was ever detected in Refs. [24, 59], which ran in approximately similar conditions. Both works attributed the excess events measured in Ta compared to W to nuclear deexcitations. In Ref. [24], the experiment was performed in 100 time series, but the integration time for each series was not specified. It is also difficult to compare the total X-ray yield. In Ref. [24], the prompt plasma radiation is said to determine a single pulse, whereas the reported total counts ($\sim 10^5$) are probably collected at delay times $t \geq 3$ µs following plasma ignition. In our study, 10^8 events were measured in a 300 s timeframe, recording all the events detected

[9] Altering the orientation of the detector to face the sample directly rather than downwards could potentially increase the counts. However, this change may also shorten the longevity of the experiment due to increased depositions.

from plasma ignition with $E_\gamma \geq 3\,\text{keV}$. Presumably, we could have obtained higher statistics, although it is difficult to make a direct comparison based on the published information. This noted, the authors assert that they have observed 49 excess events in ^{181}Ta that can be attributed to its isomer radiative decay. This leads to an average of one isomer decay detected event for every two time series.

Similarly, in Ref. [59], excess counts measured in the signal generated by delayed electrons were attributed to the depletion of the ^{181}Ta isomer. In this work, with only 8000 laser shots[10] at a repetition rate of 1 kHz, delayed electrons were measured in excess for a Ta sample that—based on their estimates—yielded to $N_{\text{exc}} \sim 30to50$ excited nuclei per laser shot. Accounting for the difference in the IC and γ-decay channels, this should have produced in our geometry and measurement window $\sim 10^5 - 10^6$ detectable events. As discussed in detail in Sect. 5.6 and particularly in Fig. 5.18, it seems that the excess counts of Ta with respect to W are not a reliable signature of the isomer decay in this scenario.

Another potential explanation for the apparent absence of signal from $^{181\text{m}}$Ta could be the low detection efficiency in measuring photons at $\sim 6\,\text{keV}$ due to its proximity to the detector threshold $D_{\text{th}} = 3\,\text{keV}$. If, for example, a photon releases its energy in multiple pixels, the energy deposited in each pixel may be lower than the threshold, resulting in lost events. Therefore, a calibration of the AdvaPix TPX3 using a ^{55}Fe source (emitting Mn X-rays, K_α at 5.9 keV and K_β at 6.5 keV)—as it was done in Ref. [24]—is necessary to estimate the detection efficiency of an eventual γ-ray emitted by the nuclear deexcitation.[11]

Of course, there is always the possibility that the scaling laws in Eq. 5.1 are overestimating the real electron temperature T_e that we can achieve in the experiment, which could make nuclear excitation even more challenging than what predicted by Tables 5.3, 5.4 and 5.5. To address this concern, we can take a closer look at the signals observed in the energy-time map and highlight two different time windows, as shown in Fig. 5.23. It can be observed that the events in the time window $t_1 = 250\,\text{ns} \leq t \leq 1000\,\text{ns}$ can be divided into two further portions: labeled as 'P1' in the interval $t_1 \leq t \leq t_2 = 450\,\text{ns}$, and 'P2' in the interval $t_2 < t \leq 1000\,\text{ns}$. The first portion P1 displays a broad energy spectrum with significant counts up to 50 keV, while P2 shows a different behavior, with the energy decreasing with time. Interestingly, this behavior is similar to that observed in the delayed portion of the signal, labeled with 'D,' in the time interval $1200\,\text{ns} \leq t \leq 3200\,\text{ns}$.

In Fig. 5.24, the spectra for the three different subsets of detected events are shown. In panel (a) of Fig. 5.24, P1 is fitted with a Planck's distribution B_λ having a temperature $T_\gamma \sim 794\,\text{eV}$ and a peak at $E_\gamma \sim 4\,\text{keV}$. Since the time response of the detector is much slower than the plasma lifetime, this spectrum has to be interpreted as an average of the plasma emission over its entire expansion. Therefore, we could

---

[10] For comparison, the data presented in Fig. 5.22 were collected over a period that encloses 1.44×10^8 laser shots.

[11] Since completion of the original thesis, this calibration has been performed and we estimated the total detection probability of the characteristic ^{55}Mn X-ray photons to be $\simeq 42\%$. This confirms the suitability of the detector for the measurement of the $^{181\text{m}}$Ta isomer decay.

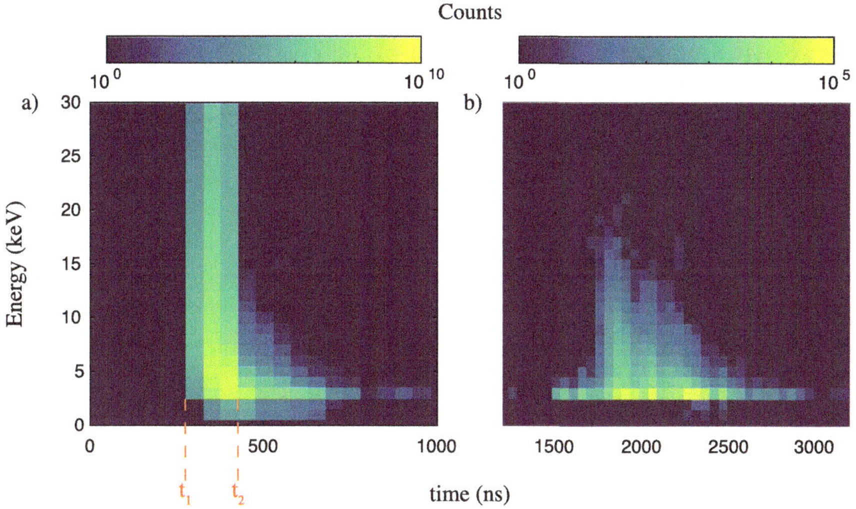

Fig. 5.23 Optical-laser generated plasma experiment: taking a closer look at the prompt and delayed signals. The 2D histogram of Fig. 5.22 has been divided into two smaller portions: in **a** the time interval $0 \leq t \leq 1000$ ns is highlighted, while in **b** $1200\,\text{ns} \leq t \leq 3200$ ns. The time instants $t_1 = 250$ ns and $t_2 = 450$ ns are indicated with vertical dashed red lines

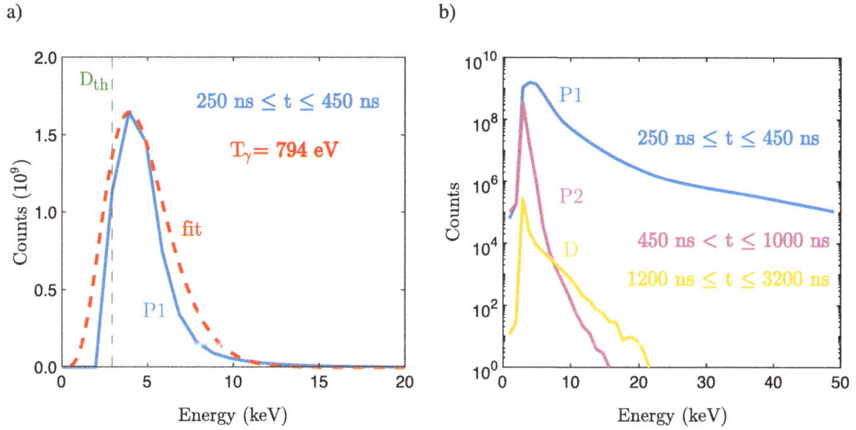

Fig. 5.24 Optical-laser generated plasma experiment: spectra of the prompt and delayed signals. Panel **a** shows the spectrum obtained by summing the events in the time intervals $250\,\text{ns} \leq t \leq 450$ ns from Fig. 5.23a. The dashed red line represents the fit obtained using Planck's B_λ distribution. The frequency parametrization B_ν results in a higher temperature $T_\gamma = 1.3$ keV, but is less accurate in fitting the data. Panel **b** displays the spectra of P1 (blue), P2 (magenta), and D (yellow) in a logarithmic scale

reasonably expect that the temperature of the photon gas T_γ is greater than 1 keV, as discussed in Sect. 5.6.

P2 could offer additional elements in support of an experimental T_e compatible with the value of 8 keV derived with SL1. In order to comment on these aspects, a comparison between the spectra of these three portions is shown in panel (b) of Fig. 5.24. Resemblances in the time dependence and energy spectra of P2 and D signals may suggest that these events share a similar origin, potentially associated with the interaction of plume ions with the deposit screen or chamber walls. As seen in Fig. 5.23, the signal P2 follows the peak of P1 by $\sim$ 100 ns. If P2 represents the hot portion of the plasma hitting the screen, this time delay is coherent with the travel time predicted by the hydrodynamic model ($\sim$ 70 ns), thus indicating that the electron temperature could be $T_e \sim 8$ keV. Additionally, signal D exhibits similar behavior to P2 but with much fewer counts. This signal could be related to the colder region of the plume reaching the deposit screen after $\sim$ 1.5 μs or to the plume interacting with the chamber walls, as discussed in the previous section. Both P2 and D have energies below 20 keV, which is consistent with the ionization of the L-shell of Ta [60]. Yet, this seems to counter the interpretation of D being a colder region of the plasma since a cooler portion would be expected to be associated with a narrower energy spectrum. The understanding of the signal D remains therefore uncertain and could be due to possible readout artifacts. As previously noted, the eventual signal of an ion striking the deposit screen while traveling at $v_{\text{exp}} = 2.63 \times 10^{-3}c$ will overlap with the prompt plasma radiation and the time response of the silicon detector, as could be the case for P2.

The discussions in this chapter have shown that the current experimental setup can reliably reproduce plasma conditions where nuclear excitation might occur. However, the absence of detected isomer decay leaves many questions open. At this stage, it is essential to identify other isotopes from those shown in Fig. 5.4 that could provide a higher excitation rate. This would enable us to study the nuclear excitations and offer a comparison with the theoretical models of the laser-generated plasma scenario.

References

1. Matinyan S (1998) Lasers as a bridge between atomic and nuclear physics. Phys Rep 298:199–249
2. Feng J et al (2022) Femtosecond pumping of nuclear isomeric states by the Coulomb collision of ions with quivering electrons. Phys Rev Lett 128:052501
3. Gibbon P, Förster E (1996) Short-pulse laser-plasma interactions. Plasma Phys Control Led Fusion 38:769
4. Krainov VP, Smirnov MB (2002) Cluster beams in the super-intense femtosecond laser pulse. Phys Rep 370:237–331
5. Löpez-Claros M, Vadillo JM, Laserna JJ (2015) Determination of plasma igni- tion threshold fluence during femtosecond single-shot laser ablation on metallic samples detected by optical emission spectroscopy. J Anal At Spectrom 30:1730–1735
6. Schmidt V, Husinsky W, Betz G (2000) Dynamics of laser desorption and ablation of metals at the threshold on the femtosecond time scale. Phys Rev Lett 85:3516

7. Harston M, Chemin J (1999) Mechanisms of nuclear excitation in plasmas. Phys Rev C 59:2462
8. Gosselin G, Morel P (2004) Enhanced nuclear level decay in hot dense plasmas. Phys Rev C 70:064603
9. Wu Y, Gunst J, Keitel CH, Pálffy A (2018) Tailoring laser-generated plasmas for efficient nuclear excitation by electron capture. Phys Rev Lett 120:052504
10. Gunst J, Wu Y, Keitel CH, Pálffy A (2018) Nuclear excitation by electron capture in optical-laser-generated plasmas. Phys Rev E 97:063205
11. Comet M et al (2015) Nuclear excitation by electron transition rate confidence interval in a Hg201 local thermodynamic equilibrium plasma. Phys Rev C 92:054609
12. Tkalya E, Akhrameev E, Arutyunyan R, Bol'Shov L, Kondratenko P (2014) Excitation of atomic nuclei in hot plasma through resonance inverse electron bridge. Phys Rev C 90:034614
13. Tkalya E, Akhrameev E, Arutyunayn R, Bol'Shov L, Kondratenko P (2012) Cross sections of electron excitation of atomic nuclei in plasma. Phys Rev C 85:044612
14. Gobet F, Dzyublik AY, Gosselin G, Méot V, Versteegen M (2022) Expected yields of ^{181}Ta(e, e)^{181}Ta* in the multi-keV range with a plasma-cathode electron beam. Phys Rev C 105:014608
15. Carroll J, Anderson J, Glesener J, Eberhard C, Collins C (1989) Accelerated decay of ^{180}Ta^{m2} and ^{176}Lu in stellar interiors through gamma, gamma-prime reactions. Astrophys J 344:454–459
16. Schumann M, Käppeler F, Böttger R, Schölermann H (1998) Coulomb excitation of 180 Ta. Phys Rev C 58:1790
17. Helmrich S (2011) Nuclear excitation by electron capture in stel lar environments. PhD thesis. Ruprecht-Karls-Universität Heidelberg, Germany
18. Letokhov V (1974) Pumping of nuclear levels by x-ray radiation of a laser plasma. Sov J Quantum Electron 3:360
19. Walker PM, Carroll JJ (2007) Nuclear isomers: recipes from the past and ingredients for the future. Nucl Phys News 17:11–15
20. Belic D et al (2002) Photo-induced depopulation of the ^{180}Ta^{m2} isomer via low-lying intermediate states: structure and astrophysical implications. Phys Rev C 65:035801
21. Izawa Y, Yamanaka C (1979) Production of ^{235}Um2 by nuclear excitation by electron transition in a laser produced uranium plasma. Phys Lett B 88:59–61
22. Goldanskii V, Namiot (1981). Excitation of nuclear isomeric levels in a hot plasma by inverse internal electron conversion. Sov J Nucl Phys(Engl. Transl.) 33
23. Arutyunyan R et al (1991) Cross section for excitation of the isomer ^{235m}U in the plasma produced by an electron beam. Sov J Nucl Phys(English Translation) 53:23–26
24. Andreev A et al (1999) Excitation of tantalum-181 nuclei in a high-temperature femtosecond laser plasma. J Exp Theor Phys Lett 69:371–376
25. Andreev AV et al (1999) Excitation of low-lying nuclear levels in a nonrelativistic hot dense laser-produced plasma. Quantum Electron 29:55
26. Borisyuk P et al (2018) Excitation of ^{229}Th nuclei in laser plasma: the energy and half-life of the low-lying isomeric state. arXiv:1804.00299
27. Arber T et al (2015) Contemporary particle-in-cell approach to laser-plasma modelling. Plasma Phys Control led Fusion 57:113001
28. Pukhov A (1999) Three-dimensional electromagnetic relativistic particle-in-cell code VLPL (Virtual Laser Plasma Lab). J Plasma Phys 61:425–433
29. Chung H-K, Chen M, Morgan W, Ralchenko Y, Lee R (2005) FLYCHK: Generalized population kinetics and spectral model for rapid spectroscopic analysis for all elements. High Energy Density Phys 1:3–12
30. Gargiulo S, Madan I, Carbone F (2022) Nuclear excitation by electron capture in excited ions. Phys Rev Lett 128:212502
31. LiveChart of Nuclides (2022) (IAEA, Nuclear data section) https://nds.iaea.org/livechart
32. Gu MF (2008) The flexible atomic code. Can J Phys 86:675–689
33. Kibedi T, Burrows T, Trzhaskovskaya MB, Davidson PM, Nestor CW Jr (2008) Evaluation of theoretical conversion coefficients using Br Icc. Nucl Instrum Methods Phys Res, Sect A 589:202–229

34. Poikela T et al (2014) Timepix3: a 65K channel hybrid pixel readout chip with si- multaneous ToA/ToT and sparse readout. J Instrum 9:C05013
35. Harilal S, Bindhu C, Tillack M, Najmabadi F, Gaeris A (2003) Internal structure and expansion dynamics of laser ablation plumes into ambient gases. J Appl Phys 93:2380–2388
36. RAEA Next-Generation Ultrafast Ti:sapphire Amplifier www.kmlabs.com/raea-ultrafast-amplifier. Accessed on 15.01.2020
37. Eliezer S (2002) The interaction of high-power lasers with plasmas. CRC Press
38. Mulser P, Bauer D (2010) High power laser-matter interaction. Springer
39. Kruer W (2019) The physics of laser plasma interactions. CRC Press
40. Denisov N (1957) On a singularity of the field of an electromagnetic wave propagated in an inhomogeneous plasma. Sov Phys JETP 4:544–553
41. Godwin R, Sachsenmaier R, Sigel R (1977) Angle-dependent reflectance of laser- produced plasmas. Phys Rev Lett 39:1198
42. Brunel F (1987) Not-so-resonant, resonant absorption. Phys Rev Lett 59:52
43. Bonnaud G, Gibbon P, Kindel J, Williams E (1991) Laser interaction with a sharp-edged overdense plasma. Laser Part Beams 9:339–354
44. Gibbon P, Bell A (1992) Collisionless absorption in sharp-edged plasmas. Phys Rev Lett 68:1535
45. Ping Y et al (2008) Absorption of short laser pulses on solid targets in the ultrarelativistic regime. Phys Rev Lett 100:085004
46. Oxenius J (1986) Kinetic theory of particles and photons. Theoretical foundations of Non-LTE plasma spectroscopy. Springer Science & Business Media
47. Pratt R et al (1977) Bremsstrahlung energy spectra from electrons of kinetic energy 1 keV $\leq$T1$\leq$2000 keV incident on neutral atoms 2 $\leq$Z$\leq$ 92. At Data Nucl Data Tables 20:175–209
48. Rybicki GB, Lightman AP (1991) Radiative processes in astrophysics. Wiley
49. Ghisellini G (2013) Radiative processes in high energy astrophysics. Springer
50. Blumberg H, Hager R, Seltzer E (1969) The internal conversion of the 6 keV transition in ^{181}Ta. Nucl Phys A 136:624–630
51. Dragoun O, Blumberg H (1971) Anomalous internal conversion of E1 transitions in outer atomic shells. Czechoslovak J Phys B 21:785–793
52. Karpeshin F, Band I, Trzhaskovskaya M (1999) 3.5-eV isomer of 229mTh how it can be produced. Nucl Phys A 654:579–596
53. Cengel Y (2002) Heat transfer: a practical approach. McGraw-Hill, New York
54. Qi J, Zhang H, Wang X (2023) Isomeric excitation of Th^{229} in laser-heated clusters. Phys Rev Lett 130:112501
55. AdvaPix TPX3 (2022). https://advacam.com [2022, October 27]
56. Jakubek J (2009) Energy-sensitive X-ray radiography and charge sharing effect in pixelated detector. Nucl Instrum Methods Phys Res, Sect A 607:192–195
57. Willmott P, Huber J (2000) Pulsed laser vaporization and deposition. Rev Mod Phys 72:315
58. Born M, Wolf E (2013) Principles of optics: electromagnetic theory of propagation, interference and diffraction of light. Elsevier
59. Savel'ev A, Chefonov O, Ovchinnikov A, Agranat M, Spohr K (2017) Direct detection of delayed high energy electrons from the 181Ta target irradiated by a moderate intensity femtosecond laser pulse. Plasma Phys Control Led Fusion 59:035004
60. Kramida A, Ralchenko Y, Reader J, NIST ASD Team (2023) NIST atomic spectra database (ver. 5.10), "physics.nist.gov/" asd. National Institute of Standards and Technology, Gaithersburg, MD, p 2022
61. Gargiulo, et al., Revisiting the excitation of the low-lying 181mTa isomer in optical laser-generated plasma. https://arxiv.org/abs/2403.11280

Chapter 6
Beyond Dichotomies: Reflections and Takeaways

Each one of us has within him a whole world of things, each man of us his own special world. And how can we ever come to an understanding if I put in the words I utter the sense and value of things as I see them; while you who listen to me must inevitably translate them according to the conception of things each one of you has within himself. We think we understand each other, but we never really do.

— Luigi Pirandello, *Six Characters in Search of an Author*

I find it fascinating how we construct our conversations and the manner in which we incorporate *logos* within them. And to be honest, I have been noticing this since I was a teenager. The moment I was able to learn and apply a more rigorous framework, rather than relying solely on intuition, was truly enlightening. Perelman and Olbrechts-Tyteca [1] write: <<*dialogue, as we consider it, is not supposed to be a debate in which the partisans of opposed settled convictions defend their respective views, but rather a discussion in which the interlocutors search honestly and without bias for the best solution to a controversial problem. Certain contemporary writers who stress this heuristic viewpoint, as against the eristic one, hold that discussion is the instrument for reaching objectively valid conclusions*>>. *Dissensus* [2] and *pro-contra*[1] arguments [3] lie at the very heart of the scientific method. We are encouraged to subject every idea to rigorous scrutiny and doubt, forge them in the *crucible of doubt*. However, this ideal is not always achieved in practice—more often than we would like to admit. As Perelman and Olbrechts-Tyteca note: <<*in a debate, on the other hand, each interlocutor advances only arguments favorable to his own thesis,*

[1] Every *pro* arguments is countered by a *contra* argument. As discussed in Ref. [3], this is the typical dichotomy with which Dostoevkij built *The Karamazov Brothers*.

S. Gargiulo, *Electromagnetic Processes of Nuclear Excitation*, Springer Theses,
https://doi.org/10.1007/978-3-031-56262-4_6

and his sole concern for arguments unfavorable to him is for the purpose of refuting them or limiting their impact. The man with a settled position is thus one-sided, and because of his bias and the consequent restriction of his efforts to those pertinent arguments that are favorable to him, the others remain frozen, as it were, and only appear in the debate if his opponent puts them forward>>. By acknowledging the importance of dialogue in our research and conversations, we can strive to create a more balanced and constructive discourse. This approach not only benefits the scientific community but also promotes the growth and development of knowledge for society as a whole [2].

As the main vision behind the manipulation of nuclear excitations is to achieve the ability to store and release energy at will, I delved into literature spanning climate science, the broader energy sector, its macroeconomic context, and potential technological innovations. In my experience participating in various panels, as well as engaging in both in-person and virtual discussions about addressing climate change, I have observed the influence of individual ideologies on the conversations (which may seem quite obvious). It is intriguing how these discussions often shift from the primary goal of addressing a global issue to defending one's rationale for choosing a particular solution or joining a specific project. It might seem naive to express this opinion, and I could be accused of such. Rather than thoughtfully examining the challenges and selecting the most effective tools to combat them, we often start and lean toward the tools that are accessible and familiar. It is this inclination that subsequently influences the problems we decide to address. In other words, our assessments of feasibility within our existing cognitive framework shape the narrative and build the connections necessary to bridge the gap between our present achievements and the issues that truly require our attention (i.e. what we should be addressing). This idea can be better illustrated using Fig. 6.1, where I have adapted[2] the representation of the structure of the *known* from Ref. [4]. This observation may offer critical insights into why science has demonstrated reduced disruptiveness across all disciplines over time, even though the number of researchers involved and the volume of literature produced have reached unprecedented levels in recent decades [5].

The question remains: how can we break this spell? Could it be beneficial to cease identifying ourselves in our current activities—that might be a construct of our training [6]—and instead concentrate on addressing the issues that truly warrant our attention and concern? We are caught in a circular definition. Do we not already believe that what we do aligns with our genuine interests and priorities?

As I look forward to the future, my aspiration is to expand and evolve in both thought and action. Cultivating self-awareness of our current thought patterns and limitations is likely the starting point that can lay the foundation for the innovations we may need. By recognizing and acknowledging our cognitive biases and constraints, we can initiate the exploration of diverse perspectives and formulate novel strategies, ultimately encouraging a more innovative and disruptive mindset. As we

[2] Panel **(a)** is adapted from Fig. 1 of Ref. [4], where it is used to depict the structure of the *known* and how we plan our actions to move from the significance of the current state of experience to what we envision as a desirable future.

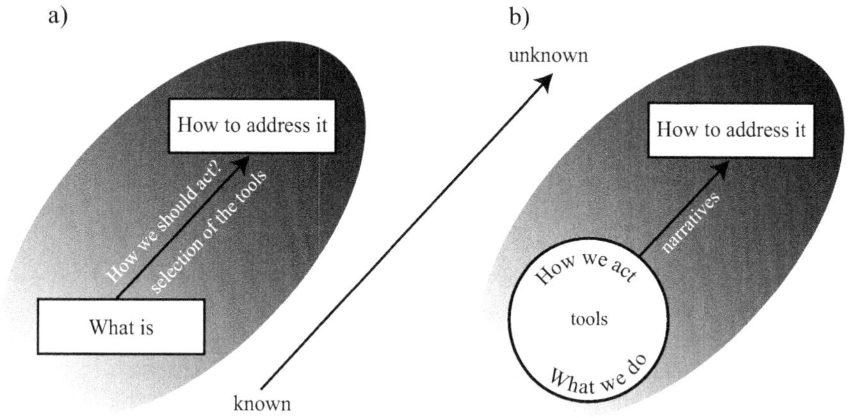

Fig. 6.1 Navigating the Unknown. Panel **a** shows the cognitive framework where we are supposed to decide the tools and the actions to be taken in order to solve the challenge, moving from the known (in white) to the unknown (black). Panel **b** illustrates a pattern of thought that may be observed in many discussions. One often begins with the tools that are somewhat accessible and proceeds to construct narratives that help bridge the gap between the current position and the challenges one aims to address. This figure is inspired by Ref. [4]

endeavor to liberate ourselves from the constraints of our beliefs, I will undertake this deconstructive exercise on myself, challenging the perspective presented in the introduction and across the thesis.

When I first became aware of the proposal to use isomers for energy storage, I was captivated, both by the inherent nature of isomerism and the remarkably long duration over which certain excitations could be maintained. The desire to harness control over nuclear decay is undeniably intriguing, as it presents a challenge against Nature that humanity may eventually win. In Chap. 2, we delved into the nature of isomers and their role as exceptional tools for probing nuclear structure, as well as their potential impact across various fields. As indicated by its title, the primary focus of this thesis was to explore the possibility of exciting nuclear levels through diverse processes, ranging from conventional photoabsorption to second-order processes like NEEC and NEμC, which involve the capture of a free electron or muon. The historical perspectives examined throughout the thesis have offered us a chance to appreciate the intricate journey of studying these subjects, while also recognizing that in certain cases there is still much to discover or—conversely—how limited our knowledge may be. This opens up numerous possibilities for substantial advancements and potential breakthroughs as we persist in investigating aspects related to nuclear reactions and their capacity to store or produce energy. Without any doubt, the interest generated by the work on NEEC [7] has sparked a resurgence of curiosity, as demonstrated by recent publications centered on nuclear excitations, which include contributions from groups that typically operate in other fields.

Nonetheless, with the current understanding, all the processes investigated continue to exhibit complexities that make them unsuitable for a practical future in storage devices. Even if the excitation probability of NEEC for ^{93m}Mo is confirmed to be nine orders of magnitude higher than theoretical predictions, its inherent inefficiency would persist, particularly when considering the challenges and costs of producing the isomeric beam. Comparable considerations apply to other scenarios, such as ion-beam traps or laser-plasma scenarios (examined in Chap. 5), although with somewhat different issues.

The nuclear excitation processes explored, partially depicted in Fig. 2.17, involve interactions between particles or ion beams and targets under highly specific conditions. They generally display a very low occurrence rate and strict energy-matching requirements due to the narrow nuclear resonance width.[3] This necessitates employing a broad energy distribution of photons or electrons to ensure that resonance can be achieved. As a result, only a small portion of these particles interact with the nucleus, ultimately leading to excitation. Thus, these processes not only result in being inefficient, but the conditions necessary for their realization make them even less appealing. In fact, the expenditure required to store a single energy block in the form of isomers significantly outweighs any potential discussions on energy convenience.

It is challenging to conceive a scenario where isomeric energy release can be initiated and halted at will without the continuous provision of external particles such as photons, electrons, or muons. With muons, for instance, nuclei predominantly undergo fission (through orbital muon capture), altering the nature of the interaction that one would like to exploit for isomer depletion. It is also the relative simplicity and self-sustainability of fission reactions in ^{235}U with thermal neutrons that make it difficult to envision an equally efficient process for energy storage through the processes shown in Fig. 2.17. Direct energy production through fission would just be substantially more convenient; moreover, energy storage would not even be necessary in that case.

In addition, public opinion has largely turned against nuclear energy in recent years. At times the term "nuclear" itself—and I have experienced this personally—is not even well-received for public outreach efforts. This kind of hostility sometimes originates also between individuals of different fields, who tend to primarily advocate for the projects they personally invest their time in. This situation made me wonder whether, in the roles we assume, we are merely defending our preferences and beliefs—as if we were members of a cult—or we are using the *logos* to really seek an objective solution.

Rather than addressing the concerns associated with nuclear fission, society has accepted its decline, slowly dismissing the entire field (at times with understandable worries). However, several recent studies suggest that nuclear energy still retains potential as a power source [8, 9], especially in transitioning to a decarbonized grid, perhaps indicating that we might not be yet ready to phase it out.

---

[3] An exception is represented by the NEEC process that captures electrons initially bound in a solid target (NETEC), which requires an isomeric beam as discussed in Chap. 3.

All of these aspects created mixed feelings in me. While, on the one hand, I enjoy the process of discovery, bringing to reality something from the realm of the unknown, I began to question whether delving deeper into the topics presented in this thesis would truly serve their practical purpose.

From my journey in the domain of nuclear excitations, I am convinced that there are still territories awaiting exploration by a creative mind. These uncharted places are of significance for fundamental physics, primarily from a theoretical standpoint. At the same time, one must also be aware of the limitations present in case one would like to aim for an immediate impact. On this front, as our understanding deepens in nuclear reactions, we may uncover new possibilities that capture interest and ultimately encourage global consensus for their advancement. Certainly, tackling climate change necessitates innovation and research, perhaps as this one, across diverse fields, since the true source of a transformative revolution might still be hidden. And this, at its core, embodies the spirit of scientific endeavor. However, it is also crucial not to depend solely on the prospect of discovering something new or extraordinary, especially given the urgency of the situation. Reevaluating and innovating[4] within existing nuclear technologies could help address concerns that currently hinder their wider adoption, providing a robust carbon-free energy foundation upon which we can pursue further discoveries in green solutions. A multidisciplinary scientific environment featuring a polyphony of opinions that may clash, intertwine, or remain distant, with the sole intent to support ideas with real potential for impact—irrespective of one's original fields of expertise or personal narratives—could provide the framework that could help us in addressing this global challenge. Embracing such a mindset can be both enlightening and liberating.

In conclusion, I hope this thesis may help new students to navigate this small domain of physics, in the way I wished I had at the beginning of my Ph.D.

References

1. Perelman C, Olbrechts-Tyteca L (1971) Traité de l'argumentation. La nouvelle rhétorique
2. Amossy R (2014) Apologie de la polémique. Presses universitaires de France Paris
3. Zagrebelsky G (2015) Liberi servi. Il Grande Inquisitore e l'enigma del potere, Torino: Einaudi
4. Peterson JB (2002) Maps of meaning: the architecture of belief. Routledge
5. Park M, Leahey E, Funk RJ (2023) Papers and patents are becoming less disruptive over time. Nature 613:138–144
6. De La Boétie E, Tournon L, Audegean P (2002) Discours de la servitude volontaire. Vrin
7. Chiara C et al (2018) Isomer depletion as experimental evidence of nuclear excitation by electron capture. Nature 554:216–218
8. Sepulveda NA, Jenkins JD, de Sisternes FJ, Lester RK (2018) The role of firm low-carbon electricity resources in deep decarbonization of power generation. Joule 2:2403–2420
9. Freese LM, Chossière GP, Eastham SD, Jenn A, Selin NE (2023) Nu- clear power generation phase-outs redistribute US air quality and climate-related mortality risk. Nat Energy 1–12

[4] There are already companies that are, for example, trying to address safety and long-lived waste issues by reshaping and 'reinventing' the concept of nuclear energy.

Appendix

A.1 I=0^{+-} Isomeric States

See Table A.1.

Table A.1 Isomer states with 0^{+-} total nuclear angular momentum. Assignments between round brackets are uncertain. Data retrieved from the LiveChart of Nuclides database [1]. LiveChart values are extracted from the ENSDF snapshot of April 2022

Isotope	I^{π}	Energy (keV)	Half-life
^{12}Be	0+	2251.0	230.0 ns
^{16}N	0−	120.4	5.3 us
^{26}Al	0+	228.3	6.3 s
^{44}S	0+	1365.0	2.6 us
^{38}K	0+	130.2	924.4 ms
^{40}K	0+	1643.6	0.3 us
^{44}Sc	0−	146.2	51.0 us
^{68}Ni	0+	1603.5[a]	270.0 ns
^{68}Ni	0+[b]	2511.9	≤15.0 ns
^{70}Ni	(0+)	1567.1	≤70.0 ns
^{72}Ga	(0+)	119.7	39.7 ms
^{74}Ga	(0+)	59.6	9.5 s
^{72}Ge	0+	691.4	444.2 ns
^{72}Se	0+	937.2	17.5 ns
^{72}Kr	0+	671.0	26.3 ns
^{74}Kr	0+	509.0	13.0 ns
^{98}Sr	0+	215.6	22.9 ns
^{90}Zr	0+	1760.7	61.3 ns
^{96}Zr	0+	1581.6	38.0 ns

(continued)

S. Gargiulo, *Electromagnetic Processes of Nuclear Excitation*, Springer Theses,
https://doi.org/10.1007/978-3-031-56262-4

Table A.1 (continued)

Isotope	I^{π}	Energy (keV)	Half-life
^{98}Zr	0+	854.1	64.0 ns
^{98}Mo	0+	734.8	21.8 ns
^{102}Pd	0+	1593.2	14.5 ns
^{118}Ag	0(−) TO 2(−)	45.8	≈ 0.1 us
^{150}Eu	0−	41.7	12.8 h
^{152}Eu	0−	45.6	9.3 h
^{156}Tb	(0+)	88.4	5.3 h
^{158}Tb	0−	110.3	10.7 s
^{166}Lu	0−	43.0	2.1 m
^{176}Ta	(0+)	100.2	30.5 ns
^{180}Ta	0−	107.8	19.2 ns
^{204}Tl	(0-)	145.9	18.7 ns
^{234}Pa	(0−)	73.92+X[c]	1.2 m
^{236}U	(0+)	2750.0	67.0 ns
^{236}U	(0+)	3434.0	≤20.0 ns
^{238}U	0+	2557.9	280.0 ns

[a]See Ref. [2]
[b]See Ref. [3]
[c]In Ref. [4] $X = 2.6\text{keV} \pm 0.5\text{keV}$

A.2 0 → 0 E0 Isomeric Transitions

See Table A.2.

Table A.2 Isomeric E0 transitions towards a I = 0 **ground state.** Data retrieved from the LiveChart of Nuclides database [1]. Assignments between round brackets are uncertain. LiveChart values are extracted from the ENSDF snapshot of April 2022

Isotope	I^{π}	Energy (keV)	Half-life
^{12}Be	0+	2251.0	230.0 ns
^{44}S	0+	1365.0	2.6 us
^{68}Ni	0+	1603.5[a]	270.0 ns
^{68}Ni	0+[b]	2511.9	≤15.0 ns
^{70}Ni	(0+)	1567.1	≤70.0 ns
^{72}Ge	0+	691.4	444.2 ns
^{72}Se	0+	937.2	17.5 ns
^{72}Kr	0+	671.0	26.3 ns
^{74}Kr	0+	509.0	13.0 ns

(continued)

Table A.2 (continued)

Isotope	I^{π}	Energy (keV)	Half-life
^{98}Sr	0+	215.6	22.9 ns
^{90}Zr	0+	1760.7	61.3 ns
^{96}Zr	0+	1581.6	38.0 ns
^{98}Zr	0+	854.1	64.0 ns
^{98}Mo	0+	734.8	21.8 ns
^{102}Pd	0+	1593.2	14.5 ns
^{236}U	(0+)	3434.0	≤20.0 ns
^{238}U	0+	2557.9	280.0 ns

[a]See Ref. [2]
[b]See Ref. [3]

A.3 Recoupling coefficients for NEEC-EXI

The following discussion is built on the results obtained with the Flexible Atomic Code (FAC) [5]. This chapter provides examples of the expressions of the recoupling coefficients between the parent and child electronic configurations corresponding to the configurations before and after the electron capture, respectively. Although all these aspects may be well known by atomic and nuclear physicists, I hope this section could be useful to gather all the information in one place and apply it to the case of NEEC. The following books have been invaluable resources for this topic [6–9], as well as Refs. [10, 11], which introduce the computer program NJFORMULA that can be used to perform the symbolic calculations of the recoupling coefficients. For a graphical interpretation of the coupling schemes in terms of a binary tree, please refer to the Supplemental Material provided in Ref. [12].

A.3.1 $N_{ele} = 0 \rightarrow N_{ele} = 1$

If one starts from a completely bare nucleus, the number of electrons in the atomic shell is equal to $N_{ele} = 0$ and the charge state $q = Z$. Therefore, $1s^0$ represents the only parent configuration possible for the electronic shell before the electron capture. In contrast, when NEEC took place in a bare nucleus, the resulting electronic configurations have $N_{ele} = 1$ and $q = Z - 1$. If one considers only capture in K, L, and M shells there are 9 possible child configurations (i.e., $1s^1_{1/2}$, $2s^1_{1/2}$, $2p^1_{1/2}$, $2p^1_{3/2}$, $3s^1_{1/2}$, $3p^1_{1/2}$, $3p^1_{3/2}$, $3d^1_{3/2}$, $3d^1_{5/2}$) for the single $1s^0$ parent configuration. All these capture channels are permitted by selection rules.

A.3.2 $\mathbf{N_{ele} = 1 \rightarrow N_{ele} = 2}$

There are 9 parent configurations and 98 child configurations. These result from all the possible combinations of electrons filling different orbitals and their consequent couplings (i.e., $1s^1_{1/2}2s^1_{1/2}(J = 0)$ and $1s^1_{1/2}2s^1_{1/2}(J = 1)$ are two distinct possibilities). Selection rules are satisfied if the spectator electron (the one not involved in the capture) preserves its subshell and the quantum number j before and after the capture.

A.3.3 $\mathbf{N_{ele} = 2 \rightarrow N_{ele} = 3}$

When moving from 2 to 3 electrons, we pass from 98 parent configurations to 665 child combinations. We need to determine which of these many possibilities are allowed by selection rules. The situation slightly changes: with three electrons there is the possibility that two electrons will fill the same shell, closing it (i.e., $s_{1/2}$, $p_{1/2}$), and leaving the remaining orbital open. It is important to keep in mind that FAC does not include in the output the electronic configurations that are closed.

Thus, if the NEEC channel is the one that closed the shell, this latter will not be formally found in the child configuration. One has to distinguish thus two cases: (i) capture shell is contained in the child configuration or (ii) not. To indicate parent and child configurations I used the following names: shell-in for the parent configuration and shell-out for the child. As a consequence, the two cases become (i) shell-out$\neq \varnothing$ or (ii) shell-out$= \varnothing$. The reason for this nomenclature is that looking for the name of the capture shell in the child configuration would result in a non-empty or empty set. This situation is depicted in Fig. A.1 for the case $\mathbf{N_{ele} = 3 \rightarrow N_{ele} = 4}$.

1. Shell-Out $\neq \varnothing$

One has to further distinguish between the cases in which the dimensions of the shell-out and shell-in differ or are equal.

1a. dim(shell-out) > *dim(shell in)*

In this condition, the capture occurs always creating a new shell, that is absent in the parent electronic configuration.

- If capture occurs in the **outer** shell, we must check whether the J of the single coupled shells and the coupling with the other shells are preserved for the spectator electrons. If both are true, the shell can be preserved.
- If **dim(shell-out)** $= \mathbf{3}$ (max possible) and the electron is not captured in the outer shell, it means that the electron capture broke the initial coupling, leading to a recoupling. The child configurations of interest are of the type $(j_c, j_2)j_3$ or

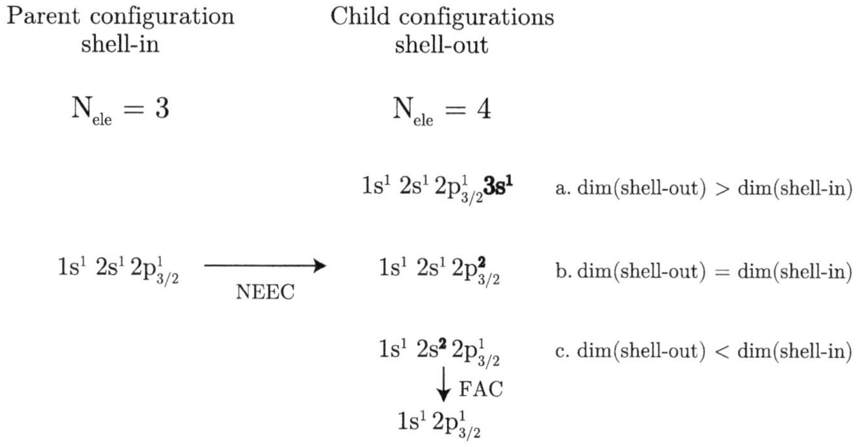

Fig. A.1 Parent and child electronic configurations for $N_{ele} = 3 \rightarrow N_{ele} = 4$. When NEEC occurs starting from the parent configuration, three different outcomes are possible depending on the dimension of shell-in and shell-out. The relativistic configuration name of the level provided by FAC includes only open and non-empty shells. Gray highlighting indicates the captured electron. This image is an exemplification and does not show all possible variations

$(j_1, j_c)j_3$. Thus the Wigner coefficients differ only in the position of j_1 and j_2. In the first case we have $j_c = j_1$ (**Case #1.1**)

$$\begin{Bmatrix} j_1 & j_2 & J_{12} \\ j_3 & J & J_{23} \end{Bmatrix}, \tag{A.1}$$

while in the second case $j_c = j_2$ (**Case #1.2**)

$$\begin{Bmatrix} j_2 & j_1 & J_{12} \\ j_3 & J & J_{13} \end{Bmatrix}, \tag{A.2}$$

where J is the total angular momentum of the child configuration.

- If **dim(shell-out) = 2** we have two possibilities (gray highlighting indicates the captured electron):
  - **Case #1.3**: $3d^2_{5/2} \rightarrow 2p^1_{1/2}\, 3d^2_{5/2}$
  - **Case #1.4**: $2p^2_{3/2} \rightarrow 2p^2_{3/2}\, 3d^1_{1/2}$.

Case #1.4 is already considered as capture in the outer shell. For **Case #1.3**: we have two spectator electrons and a captured one. We have to check that those two electrons have the same coupled J before and after the capture. Basically we are considering the two state vectors $|j_1(j_2 j_3)J_{23}; J\rangle$ and $|(j_2 j_3)J_{23}, j_1; J\rangle$. Thus, we need to verify that the two J_{23} are the same.

1b. dim(shell-out) = dim(shell-in)

If we are in this condition, it means that the shell-out configuration might be of the type $2p_{3/2}^2$ and $2p_{3/2}^3$: the capture occurs in a shell that already includes an electron. This means that in any case, the captured electron will couple first with the electron in the same shell and then with the third one. We can distinguish three different cases:

- **Case #2.1** is $2p_{3/2}^2 \rightarrow 2p_{3/2}^3$
- **Case #2.2** is $2p_{1/2}^1\, 2p_{3/2}^1 \rightarrow 2p_{1/2}^1\, 2p_{3/2}^2$
- **Case #2.3** is $2p_{3/2}^1\, 3d_{3/2}^1 \rightarrow 2p_{3/2}^2\, 3d_{3/2}^1$

Case #2.1 is a recoupling between the three electrons. In the case of $2p_{3/2}^3$, J can have only the value 3/2. Anyway, since the three electrons are indistinguishable, the channel is taken into account as happens for conventional NEEC-GSA.

Case #2.2 Here j_2 and j_3 couple first and then they couple with j_1. The recoupling coefficient is given by $\langle(12)3)|(23)1\rangle$, where $\langle\mathbf{f}|\mathbf{i}\rangle$ denotes the initial and final state vector configurations [7]. The definition of initial and final states is irrelevant as recoupling coefficients enable the passage from one scheme to another, i.e. $\langle\mathbf{i}|\mathbf{f}\rangle \equiv \langle\mathbf{f}|\mathbf{i}\rangle$ [11]. The recoupling coefficient will be related to the state vectors $\langle(j_1 j_2)J_{12}, j_3; J|(j_2 j_3)J_{23}, j_1; J\rangle$, hence the Wigner coefficient is the same as that of Eq. A.1:

$$\begin{Bmatrix} j_1 & j_2 & J_{12} \\ j_3 & J & J_{23} \end{Bmatrix} \tag{A.3}$$

This case is distinguished from the following one by the position in which the electron is captured.

Case #2.3 Here j_1 and j_2 couple first and then j_3. For this reason, the coefficient is the one given in Eq. A.2, but since the two electrons (j_1 and j_2) belong to the same shell, they also have the same j. Therefore, the two expressions in Eqs. A.1 and A.2 are equivalent. The recoupling coefficient is given by $\langle(12)3|(13)2\rangle$:

$$\begin{Bmatrix} j_2 & j_1 & J_{12} \\ j_3 & J & J_{13} \end{Bmatrix} \tag{A.4}$$

c. dim(shell-out) < dim(shell-in)

This scenario corresponds to the case in which one shell has been closed due to the capture of a free electron. Specifically, given the presence of three electrons, it is possible that either the s, $p_{1/2}$, or $d_{3/2}$ shell could have been closed. By construction, FAC will not include this shell in the output file. Nonetheless, the search result did not give an empty set. When searching for the capture channel, I use the notation of the LS coupling (i.e., $2p$ or $3d$). Having shell-out $\neq \varnothing$ suggests, for example, the closure of the $2p_{1/2}$ subshell while the other electron is in the $2p_{3/2}$ orbital. Having a closed subshell, we have additional information on the coupled momentum of the two electrons ($J_{ij} = 0$). There is only one case (equivalent to the one involving the d shell)

- **Case #3.1** is $2p^1_{1/2}\,2p^1_{3/2} \rightarrow 2p^2_{1/2}\,2p^1_{3/2}$

Case #3.1 Here, we have something analogous to Eq. A.4, $\langle(12)3|(13)2\rangle$, with $J_{13} = 0$ and $j_1 = j_3$

$$\begin{Bmatrix} j_2 & j_1 & J_{12} \\ j_3 & J & 0 \end{Bmatrix} \tag{A.5}$$

2. shell-out $= \varnothing$

The case shell-out $= \varnothing$ implies that the search of the capture shell in the output electronic configuration did not produce any result. For example, this could occur when searching for capture in the $2p$ shell, with no $2p_{1/2}$ or $2p_{3/2}$ electron found. This means that the capture shell existed in the initial configuration and it was subsequently closed with the capture. Only the case 'c' of Fig. A.1 occurs.

2c. Dimension shell out < shell in.

We can distinguish two cases:

- **Case #4.1** $2s^1_{1/2}\,2p^1_{1/2} \rightarrow 2s^2_{1/2}\,2p^1_{1/2}$
- **Case #4.2** is $2s^1_{1/2}\,2p^1_{1/2} \rightarrow 2s^1_{1/2}\,2p^2_{1/2}$

Case #4.1 is identical to Case #3.1:

$$\begin{Bmatrix} j_2 & j_1 & J_{12} \\ j_3 & J & 0 \end{Bmatrix} \tag{A.6}$$

Case #4.2 is identical to Case #2.2 with $J_{23} = 0$:

$$\begin{Bmatrix} j_1 & j_2 & J_f \\ j_3 & J & 0 \end{Bmatrix} \tag{A.7}$$

A.3.4 $\mathbf{N_{ele} = 3 \rightarrow N_{ele} = 4}$

Here we pass from 665 parent configurations to 3793 child configurations. As similarly done for the previous cases, it is necessary to differentiate the situation in which the search of the capture shell in the output electronic configuration yielded a non-empty set.

1. shell-out $\neq \varnothing$

1a. dim(shell-out) > dim(shell-in)
In this scenario—regardless of the number of subshells present in the output configuration—the capture results in the creation of a new subshell that was absent in the parent configuration.

- If capture occurs in the **outer** shell, it is necessary to verify that the couplings of the J_{ij} between the subshells of the spectator electrons have been preserved. This is the case in which the capture of the free electron does not break the initial electron coupling: **Case #5.0**.

If the capture takes place in **inner** shells, this breaks the initial angular momentum coupling. The outcome has to be differentiated based on the size of the output configuration: if **dim(shell-out)**$=4$ the four electrons are all in different subshells. Three cases can be distinguished, depending on the position of the capture shell (indicated in blue):

- **Case #5.1** $\langle[(1,2),3],4|[(2,3),4],1\rangle$. The recoupling coefficient is reported in Eq. A.8.
- **Case #5.2** $\langle[(1,2),3],4|[(1,3),4],2\rangle$. The recoupling coefficient is reported in Eq. A.9.
- **Case #5.3** $\langle[(1,2),3],4|[(1,2),4],3\rangle$. The recoupling coefficient is reported in Eq. A.10.

if **dim(shell-out)** $=$ **3** the capture always creates a new shell, while the other two spectator electrons (out of four) are in the same shell. Furthermore, since shell-out $\neq \varnothing$, there are no shells that could have potentially been closed. As a first step, it is important to determine the location of the shell containing two electrons in the final electronic configuration. Two cases are possible:

- Case #5.4: $2p^1_{1/2}\,3d^2_{5/2}$
- Case #5.5: $2p^2_{3/2}\,3d^1_{5/2}$

Case #5.4 can be further split in

- **Case #5.4.1**: $2p^1_{1/2}\,3d^2_{5/2} \rightarrow 1s^1\,2p^1_{1/2}\,3d^2_{5/2}$. The recoupling coefficient is reported in Eq. A.11.
- **Case #5.4.2**: $2p^1_{1/2}\,3d^2_{5/2} \rightarrow 2p^1_{1/2}\,2p^1_{3/2}\,3d^2_{5/2}$. The recoupling coefficient is reported in Eq. A.12.

Case #5.5 can be further divided in

- **Case #5.5.1**: $2p^2_{3/2}\,3d^1_{5/2} \rightarrow 1s^1\,2p^2_{3/2}\,3d^1_{5/2}$. Eq. A.13.
- **Case #5.5.2**: $2p^2_{3/2}\,3d^1_{5/2} \rightarrow 2p^2_{3/2}\,3d^1_{3/2}\,3d^1_{5/2}$. Eq. A.14.

if **dim(shell-out)** $=2$ there are two cases:

- **Case #5.6**: $3d^x_{5/2} \rightarrow 1s^1\,3d^x_{5/2}$, where x can be 1 or 3.
- **Case #5.7** $2p^x_{3/2} \rightarrow 2p^x_{3/2}\,3d^1_{3/2}$, where x can be 1 or 3.

The capture in the outer shell (**Case #5.7**) is already considered under Case #5.0. For **Case #5.6**, as discussed for Case #1.3, it is necessary to check that the spectator electrons have preserved their own j.

1b. dim(shell-out) = dim(shell-in)

We can differentiate the following cases based on the size of the output shell (which can have a maximum size of 3 due to the dimension of the parent shell) and the index of the captured shell.

- **Case #6.1** is $3d^{x}_{5/2} \to 3d^{(x+1)}_{5/2}$, Accepted $\forall x$.
- **Case #6.2.1** is $2p^{2}_{3/2}\, 3d^{1}_{5/2} \to 2p^{2}_{3/2}\, 3d^{2}_{5/2}$, Eq. A.15.
- **Case #6.2.2** is $2p^{2}_{3/2}\, 3d^{1}_{5/2} \to 2p^{3}_{3/2}\, 3d^{1}_{5/2}$ Eq. A.14.
- **Case #6.2.3** is $2p^{1}_{3/2}\, 3d^{2}_{5/2} \to 2p^{2}_{3/2}\, 3d^{2}_{5/2}$ Eq. A.16.
- **Case #6.2.4** is $2p^{1}_{3/2}\, 3d^{2}_{5/2} \to 2p^{1}_{3/2}\, 3d^{3}_{5/2}$ Eq. A.17.
- **Case #6.3.1** is $1s^{1}\, 2p^{1}_{3/2}\, 3d^{1}_{5/2} \to 1s^{1}\, 2p^{1}_{3/2}\, 3d^{2}_{5/2}$ Eq. A.15.
- **Case #6.3.2** is $1s^{1}\, 2p^{1}_{3/2}\, 3d^{1}_{5/2} \to 1s^{1}\, 2p^{2}_{3/2}\, 3d^{1}_{5/2}$ Eq. A.18.
- **Case #6.3.3** $2p^{1}_{3/2} 3d^{1}_{3/2} 3d^{1}_{5/2} \to 2p^{2}_{3/2}\, 3d^{1}_{3/2}\, 3d^{1}_{5/2}$ Eq. A.19.

1c. dim(shell-out) < dim(shell-in)

We can distinguish different cases depending on the relative position of the closed shell and the size of the input shell. When **dim(shell-in)** = **3**, there are two possibilities to consider (keep in mind that the output will have a closed shell, so **dim(shell-out)** = **2**):

- **Case # 7.1.1** $1s^{1}\, 2p^{1}_{1/2}\, 3d^{1}_{3/2} \to 1s^{1}\, 2p^{1}_{1/2}\, 3s^{2}$
- **Case # 7.1.2** $1s^{1}\, 2p^{1}_{1/2}\, 3d^{1}_{3/2} \to 1s^{1}\, 2p^{2}_{1/2}\, 3s^{1}$
- **Case # 7.1.3** $1s^{1}\, 2p^{1}_{1/2}\, 3d^{1}_{3/2} \to 1s^{2}\, 2p^{1}_{1/2}\, 3s^{1}$

When **dim(shell-in)** = **2**, there are two possibilities to consider (note that the output will have a closed shell, so **dim(shell-out)** = **1**):

- **Case # 7.2.1** $2p^{2}_{3/2}\, 3d^{1}_{3/2} \to 2p^{2}_{3/2}\, 3d^{2}_{3/2}$
- **Case # 7.2.2** $1s^{1}\, 2p^{2}_{3/2} \to 1s^{2}\, 2p^{2}_{3/2}$

2. shell-out = $\varnothing$

2c. dim(shell-out) < dim(shell-in)
Similar to section "1b. dim(shell-out) = dim(shell-in)".

Complete list of the recoupling terms for $N_{ele} = 3 \to N_{ele} = 4$

The values of θ are irrelevant for our purposes since we are focused on evaluating the probability $\Lambda = |\langle \mathbf{a}|\mathbf{b}\rangle|^2$, where **a** and **b** represent the two state vectors describing the coupling (Figs. A.2, A.3, A.4, A.5 and A.6).

Case #5.1:

$$\langle [(j_1, j_2)J_{12}, j_3]J_{123}, j_4; J \,|\, [(j_2, j_3)J_{23}, j_4]J_{234}, j_1; J\rangle =$$
$$= (-1)^{\theta_1} R_1 \begin{Bmatrix} j_1 & j_2 & J_{12} \\ j_3 & J_{123} & J_{23} \end{Bmatrix} \begin{Bmatrix} j_1 & J_{23} & J_{123} \\ j_4 & J & J_{234} \end{Bmatrix}, \quad \text{(A.8)}$$
$$R_1 = \sqrt{(2 \cdot j_{12} + 1)(2 \cdot J_{123} + 1)(2 \cdot J_{23} + 1)(2 \cdot J_{234} + 1)},$$

Case #5.2:

$$\langle [(j_1, j_2)J_{12}, j_3]J_{123}, j_4; J \,|\, [(j_1, j_3)J_{13}, j_4]J_{134}, j_2; J\rangle =$$
$$(-1)^{\theta_2} R_2 \begin{Bmatrix} j_2 & j_1 & J_{12} \\ j_3 & J_{123} & J_{13} \end{Bmatrix} \begin{Bmatrix} j_2 & J_{13} & J_{123} \\ j_4 & J & J_{134} \end{Bmatrix}, \quad \text{(A.9)}$$
$$R_2 = \sqrt{(2 \cdot J_{12} + 1)(2 \cdot J_{123} + 1)(2 \cdot J_{13} + 1)(2 \cdot J_{134} + 1)},$$

Case #5.3:

$$\langle [(j_1, j_2)J_{12}, j_3]J_{123}, j_4; J \,|\, [(j_1, j_2)J_{12}, j_4]J_{124}, j_3; J\rangle =$$
$$(-1)^{\theta_3} R_3 \begin{Bmatrix} j_3 & J_{12} & J_{123} \\ j_4 & J & J_{124} \end{Bmatrix}, \quad \text{(A.10)}$$
$$R_3 = \sqrt{(2 \cdot J_{123} + 1)(2 \cdot J_{124} + 1)},$$

Case #5.4.1: $2p^1_{1/2}\, 3d^2_{5/2} \to 1s^1\, 2p^1_{1/2}\, 3d^2_{5/2}$ corresponds to the recoupling coefficient $\langle((12), (34))|((2, (34)), 1)\rangle$:

$$\langle [(j_1, j_2)J_{12}, (j_3, j_4)J_{34}]; J \,|\, [j_2, (j_3, j_4)J_{34}]J_{234}, j_1; J\rangle =$$
$$(-1)^{\theta_4} R_4 \begin{Bmatrix} j_1 & j_2 & J_{12} \\ J_{34} & J & J_{234} \end{Bmatrix}, \quad \text{(A.11)}$$
$$R_4 = \sqrt{(2 \cdot J_{12} + 1)(2 \cdot J_{234} + 1)},$$

Case #5.4.2: $2p^1_{1/2}\, 3d^2_{5/2} \to 2p^1_{1/2}\, 2p^1_{3/2}\, 3d^2_{5/2}$ corresponds to the recoupling coefficient $\langle((12), (34))|((1, (34)), 2)\rangle$:

$$\langle [(j_1, j_2)J_{12}, (j_3, j_4)J_{34}]; J \,|\, [j_1, (j_3, j_4)J_{34}]J_{134}, j_2; J\rangle =$$
$$(-1)^{\theta_5} R_5 \begin{Bmatrix} j_2 & j_1 & J_{12} \\ J_{34} & J & J_{134} \end{Bmatrix}, \quad \text{(A.12)}$$
$$R_5 = \sqrt{(2 \cdot J_{12} + 1)(2 \cdot J_{134} + 1)},$$

Case #5.5.1: $2p_{3/2}^2\, 3d_{5/2}^1 \to 1s^1\, 2p_{3/2}^2\, 3d_{5/2}^1$ corresponds to the recoupling coefficient $\langle((1,(23)),4)|(((23),4),1)\rangle$:

$$\langle [j_1,(j_2,j_3)J_{23}]J_{123}, j_4; J \,\|[(j_2,j_3)J_{23}, j_4]J_{234}, j_1; J\rangle = (-1)^{\theta_6} R_6 \begin{Bmatrix} j_1 & J_{23} & J_{123} \\ j_4 & J & J_{234} \end{Bmatrix}, \quad R_6 = \sqrt{(2\cdot J_{123}+1)(2\cdot J_{234}+1)}\,, \tag{A.13}$$

Case #5.5.2: $2p_{3/2}^2\, 3d_{5/2}^1 \to 2p_{3/2}^2\, 3d_{3/2}^1\, 3d_{5/2}^1$ corresponds to the recoupling coefficient $\langle(((12),3),4)|(((12),4),3)\rangle$:

$$\langle [(j_1,j_2)J_{12}, j_3]J_{123}, j_4; J \,\|[(j_1,j_2)J_{12}, j_4]J_{124}, j_3; J\rangle = (-1)^{\theta_7} R_7 \begin{Bmatrix} j_3 & J_{12} & J_{123} \\ j_4 & J & J_{124} \end{Bmatrix}, \quad R_7 = \sqrt{(2\cdot J_{124}+1)(2\cdot J_{123}+1)}\,, \tag{A.14}$$

Case #6.2.1: $2p_{3/2}^2\, 3d_{5/2}^1 \to 2p_{3/2}^2\, 3d_{5/2}^2$ corresponds to the recoupling coefficient $\langle((12),(34))|(((12),3),4)\rangle$:

$$\langle [(j_1,j_2)J_{12}, (j_3,j_4)J_{34}]; J \,\|[(j_1,j_2)J_{12}, j_3]J_{123}, j_4; J\rangle = (-1)^{\theta_8} R_8 \begin{Bmatrix} j_4 & j_3 & J_{34} \\ J_{12} & J & J_{123} \end{Bmatrix}, \quad R_8 = \sqrt{(2\cdot J_{34}+1)(2\cdot J_{123}+1)}\,, \tag{A.15}$$

Case #6.2.2 is $2p_{3/2}^2\, 3d_{5/2}^1 \to 2p_{3/2}^3\, 3d_{5/2}^1$ corresponds to the recoupling coefficient $\langle(((12),3)4)|(((12),4),3)\rangle$, that is identical to the Case #5.5.2.

Case #6.2.3 is $2p_{3/2}^1\, 3d_{5/2}^2 \to 2p_{3/2}^2\, 3d_{5/2}^2$ corresponds to the recoupling coefficient $\langle((12),(34))|((1,(34)),2)\rangle$:

$$\langle [(j_1,j_2)J_{12}, (j_3,j_4)J_{34}]; J \,\|[j_1,(j_3,j_4)J_{34}]J_{134}, j_2; J\rangle = (-1)^{\theta_9} R_9 \begin{Bmatrix} j_2 & j_1 & J_{12} \\ J_{34} & J & J_{134} \end{Bmatrix}, \quad R_9 = \sqrt{(2\cdot J_{12}+1)(2\cdot J_{134}+1)}\,, \tag{A.16}$$

Case #6.2.4 is $2p_{3/2}^1\, 3d_{5/2}^2 \to 2p_{3/2}^1\, 3d_{5/2}^3$ corresponds to the recoupling coefficient $\langle(1,((23),4))|((1,(23)),4)\rangle$:

$$\langle j_1,[(j_2,j_3)J_{23},j_4]J_{234};J\,|[j_1,(j_2,j_3)J_{23}]J_{123},j_4;J\rangle = (-1)^{\theta_{10}} R_{10} \begin{Bmatrix} j_4 & J_{23} & J_{234} \\ j_1 & J & J_{123} \end{Bmatrix}, \tag{A.17}$$

$$R_{10} = \sqrt{(2\cdot J_{234}+1)(2\cdot J_{123}+1)}\,,$$

Case #6.3.1: $1s^1\, 2p_{3/2}^1\, 3d_{5/2}^1 \to 1s^1\, 2p_{3/2}^1\, 3d_{5/2}^2$ corresponds to the recoupling coefficient $\langle((12),(34))|(((12),3),4)\rangle$, that is identical to Eq. A.15.

Case #6.3.2: $1s^1\, 2p_{3/2}^1\, 3d_{5/2}^1 \to 1s^1\, 2p_{3/2}^2\, 3d_{5/2}^1$ corresponds to the recoupling coefficient $\langle(((1,(23)),4)|(((12),4),3)\rangle$:

$$\langle[j_1,(j_2,j_3)J_{23}]J_{123},j_4;J\,|[(j_1,j_2)J_{12},j_4]J_{124},j_3;J\rangle = (-1)^{\theta_{11}} R_{11} \begin{Bmatrix} j_3 & j_2 & J_{23} \\ j_1 & J_{123} & J_{12} \end{Bmatrix} \begin{Bmatrix} j_3 & J_{12} & J_{123} \\ j_4 & J & J_{124} \end{Bmatrix}, \tag{A.18}$$

$$R_{11} = \sqrt{(2\cdot J_{23}+1)(2\cdot J_{123}+1)(2\cdot J_{12}+1)(2\cdot J_{124}+1)}\,,$$

Case #6.3.3 $2p_{3/2}^1 3d_{3/2}^1 3d_{5/2}^1 \to 2p_{3/2}^2\, 3d_{3/2}^1\, 3d_{5/2}^1$ corresponds to the recoupling coefficient $\langle((((12),3),4)|(((13),4),2)\rangle$:

$$\langle[(j_1,j_2)J_{12},j_3]J_{123},j_4;J\,|[(j_1,j_3)J_{13},j_4]J_{134},j_2;J\rangle = (-1)^{\theta_{12}} R_{12} \begin{Bmatrix} j_2 & j_1 & J_{12} \\ j_3 & J_{123} & J_{13} \end{Bmatrix} \begin{Bmatrix} j_2 & J_{13} & J_{123} \\ j_4 & J & J_{134} \end{Bmatrix}, \tag{A.19}$$

$$R_{12} = \sqrt{(2\cdot J_{12}+1)(2\cdot J_{123}+1)(2\cdot J_{13}+1)(2\cdot J_{134}+1)}\,,$$

Case # 7.1.1 $1s^1\; 2p_{1/2}^1\; 3d_{3/2}^1 \to 1s^1\; 2p_{1/2}^1\; 3d_{3/2}^2$ corresponds to the recoupling coefficient $\langle((12),(34))|(((12),3),4)\rangle$, where $J_{34}=0$:

$$\langle[(j_1,j_2)J_{12},(j_3,j_4)J_{34}];J\,|[(j_1,j_2)J_{12},j_3]J_{123},j_4;J\rangle = (-1)^{\theta_{13}} R_{13} \begin{Bmatrix} j_4 & j_3 & J_{34} \\ J_{12} & J & J_{123} \end{Bmatrix} \propto \begin{Bmatrix} j_4 & j_3 & 0 \\ J_{12} & J & J_{123} \end{Bmatrix}, \tag{A.20}$$

$$R_{13} = \sqrt{(2\cdot J_{34}+1)(2\cdot J_{123}+1)} = \sqrt{(2\cdot J_{123}+1)}$$

Case # 7.1.2 $1s^1\ 2p^1_{1/2}\ 3d^1_{3/2} \rightarrow 1s^1\ 2p^2_{1/2}\ 3d^1_{3/2}$ corresponds to the recoupling coefficient $\langle((1,(23)),4)|(((12),4),3)\rangle$, where $J_{23}=0$:

$$\langle[j_1,(j_2,j_3)J_{23}]J_{123},j_4;J\|[(j_1,j_2)J_{12},j_4]J_{124},j_3;J\rangle =$$
$$(-1)^{\theta_{14}}R_{14}\begin{Bmatrix} j_3 & j_2 & \cancel{J_{23}}^0 \\ j_1 & J_{123} & J_{12}\end{Bmatrix}\begin{Bmatrix} j_3 & J_{12} & J_{123} \\ j_4 & J & J_{124}\end{Bmatrix}, \quad \text{(A.21)}$$

$$\begin{aligned} R_{14} =& \sqrt{(2\cdot J_{23}+1)(2\cdot J_{123}+1)(2\cdot J_{12}+1)(2\cdot J_{124}+1)} \\ =& \sqrt{(2\cdot J_{123}+1)(2\cdot J_{12}+1)(2\cdot J_{124}+1)}\,, \end{aligned}$$

Case # 7.1.3 $1s^1\ 2p^1_{1/2}\ 3d^1_{3/2} \rightarrow 1s^2\ 2p^1_{1/2}\ 3d^1_{3/2}$ corresponds to the recoupling coefficient $\langle(((12),3),4)|(((13),4),2)\rangle$, where $J_{12}=0$:

$$\langle[(j_1,j_2)J_{12},j_3]J_{123},j_4;J\|[(j_1,j_3)J_{13},j_4]J_{134},j_2;J\rangle =$$
$$(-1)^{\theta_{15}}R_{15}\begin{Bmatrix} j_2 & j_1 & \cancel{J_{12}}^0 \\ j_3 & J_{123} & J_{13}\end{Bmatrix}\begin{Bmatrix} j_2 & J_{13} & J_{123} \\ j_4 & J & J_{134}\end{Bmatrix}, \quad \text{(A.22)}$$

$$\begin{aligned} R_{15} =& \sqrt{(2\cdot J_{12}+1)(2\cdot J_{123}+1)(2\cdot J_{13}+1)(2\cdot J_{134}+1)} \\ =& \sqrt{(2\cdot J_{123}+1)(2\cdot J_{13}+1)(2\cdot J_{134}+1)}\,, \end{aligned}$$

Case # 7.2.1 $2p^2_{3/2}\ 3s^1 \rightarrow 2p^2_{3/2}\ 3s^2$ corresponds to the $\langle((12),(34))|(((12),3),4)\rangle$ recoupling coefficient , where $J_{34}=0$. It is equivalent to Case #7.1.1 or #6.2.1.
Case # 7.2.2 $1s^1\ 2p^2_{3/2} \rightarrow 1s^2\ 2p^2_{3/2}$ corresponds to the $\langle((12),(34))|((1,(34)),2)\rangle$ recoupling coefficient, where $J_{12}=0$. It is equivalent to Case #6.2.3:

$$\langle[(j_1,j_2)J_{12},(j_3,j_4)J_{34}];J\|[j_1,(j_3,j_4)J_{34}]J_{134},j_2;J\rangle =$$
$$(-1)^{\theta_{15}}R_{15}\begin{Bmatrix} j_2 & j_1 & \cancel{J_{12}}^0 \\ J_{34} & J & J_{134}\end{Bmatrix}, \quad \text{(A.23)}$$

$$R_{15} = \sqrt{(2\cdot J_{134}+1)}\,.$$

A.4 Additional Figures

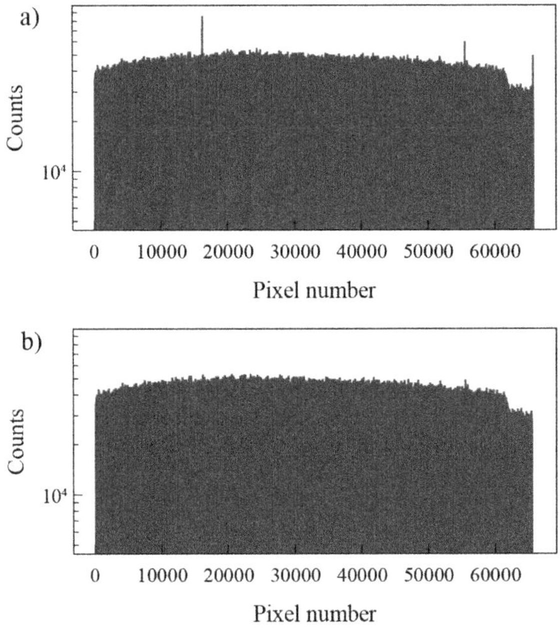

Fig. A.2 Histograms illustrating the detected event distribution as a function of pixel number, both before (a) and after (b) removing outlier pixels. Pixels are numerated from 0 to 65535. Pixels that exhibit significant deviations from the surrounding are removed before proceeding with the analysis. The slightly lower counts observed for higher pixel numbers may be attributed to the detector inclination with respect to the sample, as this feature disappears when the detector is mounted vertically and pointed directly at the sample. The analysis was alternatively conducted without including the pixels exhibiting lower count rates, but no appreciable difference was observed. To handle the large amount of data, the analysis is performed on smaller subsets of the dataset

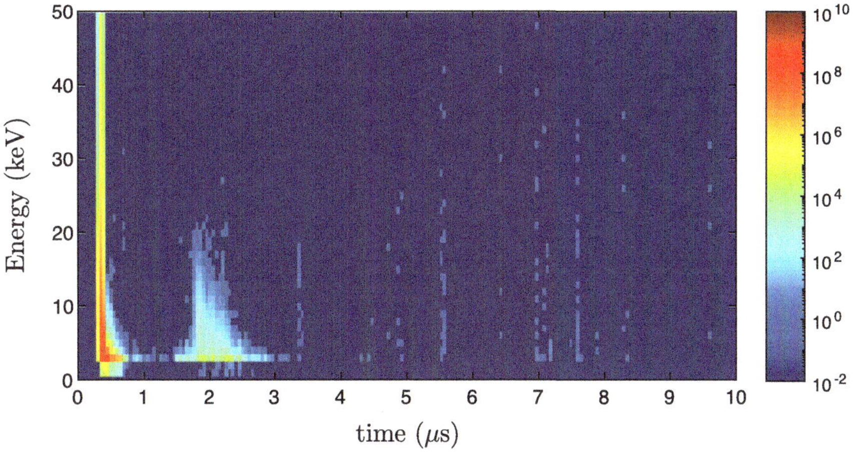

Fig. A.3 Optical-laser generated plasma experiment: 2D energy-time map. This image replicates Fig. 5.22 using a different colormap (jet) and a color axis with different limits (starting from 10^{-2} instead of 10^{0}). As a result, even single counts are now clearly visible

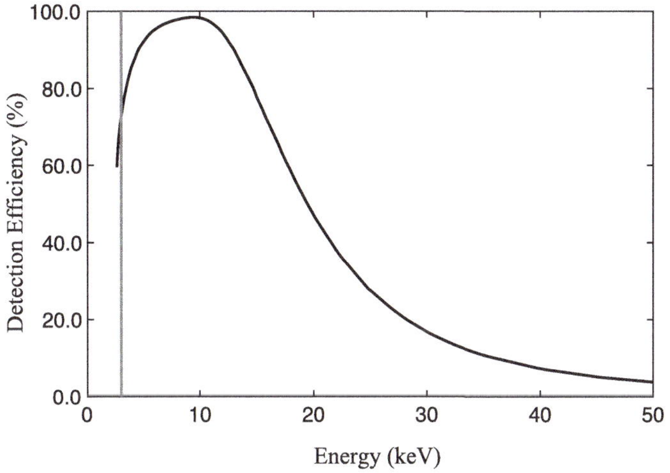

Fig. A.4 Detection efficiency of the AdvaPIX TPX3. Detection efficiency as function of energy, with the $D_{\mathrm{th}} = 3\,\mathrm{keV}$ threshold indicated by the vertical line

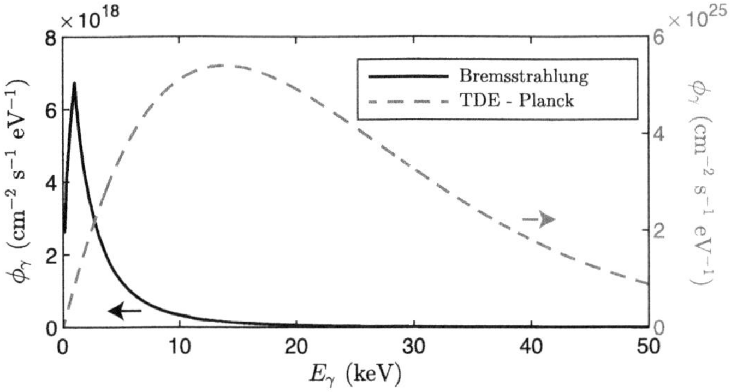

Fig. A.5 Comparison of bremsstrahlung and TDE photon flux models in a laser-generated plasma scenario. The photon fluxes ϕ_γ are determined using Eqs. 5.7 and 5.8 based on the conditions outlined by SL1. For Eq. 5.7, the value $\epsilon = 2.7 \times 10^{-6}$ is determined imposing the energy balance between the photon and electron gases

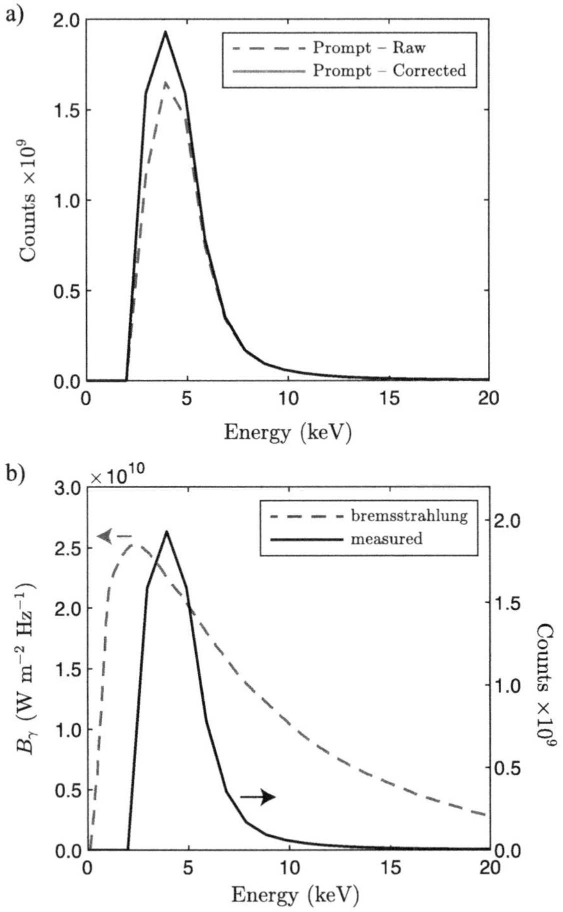

Fig. A.6 Prompt spectrum corrected by the energy-dependent detection efficiency (a) and comparison of the measured spectrum with the bremsstrahlung photon distribution (b). In **a**, the prompt raw spectrum 'P1' is corrected with the detection efficiency of the AdvaPIX TPX3. In **b**, the corrected spectrum is compared with the bremsstrahlung photon distribution, obtained from Eq. 5.8 as $B_\nu^{\mathrm{B}} = (hE/4\pi)\phi_\gamma^{\mathrm{B}}$ [13]

References

1. LiveChart of Nuclides (2022) (IAEA, Nuclear data section). https://nds.iaea.org/livechart
2. Recchia F et al (2013) Configuration mixing and relative transition rates between low-spin states in 68Ni. Phys Rev C 88:041302
3. Chiara C et al (2012) Low-spin states and the non-observation of a proposed 2202-keV, 0+ isomer in 68Ni. Phys Rev C 86:041304
4. Rimskii-Korsakov A, Koltsov V, Karasev V (2016) Scheme of the??-decay of 234 Th. Bull Russ Acad Sci Phys 80:880–883
5. Gu MF (2008) The flexible atomic code. Can J Phys 86:675–689
6. Messiah A (2014) Quantum mechanics. Courier Corporation
7. Biedenharn LC, Louck JD (1981) The Racah-Wigner algebra in quantum theory. Advanced book program. Addison-Wesley
8. Varshalovich DA, Moskalev AN, Khersonskii VK (1988) Quantum theory of angular momentum. World Scientific
9. Edmonds AR (1957) Angular momentum in quantum mechanics. Princeton University Press
10. Fack V, Pitre S, Van der Jeugt J (1995) New efficient programs to calculate general recoupling coefficients. Part II: evaluation of a summation formula. Comput Phys Commun 86:105–122
11. Fack V, Lievens S, Van der Jeugt J (1999) On rotation distance between binary coupling trees and applications for 3nj-coefficients. Comput Phys Commun 119:99–114
12. Gargiulo S, Madan I, Carbone F (2022) Nuclear excitation by electron capture in excited ions. Phys Rev Lett 128:212502
13. Oxenius J (1986) Kinetic theory of particles and photons. Theoretical foundations of Non-LTE plasma spectroscopy. Springer Science & Business Media

Curriculum Vitae

Simone Gargiulo

Nationality: Italian
Birth: April 10th 1992, Torre del Greco, Napoli, Italy
E-mail: `simone.gargiulo@epfl.ch`
`simonegargiulo2@gmail.com`

Education

April 2023	**Ph.D., Physics** École Polytechnique Fédérale de Lausanne, Switzerland Topics: nuclear isomers, nuclear excitation by electron capture (NEEC), laser-plasma physics, ultrafast electron microscopy, plasmonics. Title: *Electromagnetic processes of nuclear excitation: from direct photoabsorption to free electron and muon capture.* Group: Laboratory for ultrafast microscopy and electron scattering (LUMES). Thesis supervisor: Fabrizio Carbone.
Sept. 2018	**M.Sc., Electronic Engineering** Università degli Studi di Napoli Federico II, Italy Topics: crystal-assisted collimation in particle accelerators, X-ray diffraction from bent crystals. Title: *Modelling for metrological analysis of an X-ray diffractometer for bent crystals characterization in crystal-assisted collimation.* Final mark: 110/110 cum laude, GPA: 4.0/4.0

S. Gargiulo, *Electromagnetic Processes of Nuclear Excitation*, Springer Theses,
https://doi.org/10.1007/978-3-031-56262-4

Dec. 2014 **B.Sc., Electronic Engineering** Università degli Studi di Napoli Federico II, Italy
Topic: drug delivery systems.
Title: *Microgel Photonics.*
Final mark: 110/110 cum laude, GPA: 3.9/4.0

Sept. 2010 **High School** Istituto Tecnico Industriale Antonio Pacinotti, Scafati, Italy
Final mark: 96/100

Professional Experience

Mar. 2019–July 2023 **Researcher** Swiss Federal Institute of Technology (EPFL), Lausanne, Switzerland

- Seeking innovative solutions to address the climate crisis, Google Inc. sponsored my doctoral research, which focused on exploring groundbreaking energy harvesting or storing tools.
- Developed numerical and simulation tools for theoretical calculations related to nuclear excitation processes.
- Designed and implemented a tabletop experimental setup to investigate the excitation of nuclear levels in a neutral plasma, generated by a femtosecond laser pulse.
- Developed finite element method (FEM) models using COMSOL and LUMERICAL, covering various fields of physics, including picosecond thermal dynamics, plasmonics, and electromagnetic wave propagation.
- Particle-in-cell simulations using EPOCH and VLPL for laser-plasma interactions.

Sept. 2017–Oct. 2018 **Electronic Engineer** European Organization for Nuclear Research (CERN), Geneva, Switzerland

- Designed and commissioned high-precision instrumentation for the characterization of crystal collimators in the Large Hadron Collider (LHC).
- Developed theoretical frameworks and simulation tools for predicting experimental outcomes, accounting for real-world conditions and uncertainties.
- Developed numerical tools using COMSOL, integrating aspects of ray optics and mechanical stress-induced deformations to analyze the performance of bent crystals used as collimators in the LHC.

Jan. 2012–Feb. 2012 **Technical Education Teacher** High School Maria Montessori-L. Da Vinci, Porretta Terme, Italy
• Delivered theoretical lectures and mentored students in the experimental design of electronic circuits.
• Assessed student progress to evaluate the effectiveness and impact of teaching methods and assigned projects.
• Supervised four classes consisting of 20 students each.

Research Projects (selected)

March 2019–April 2023 **Processes of Nuclear Excitation** EPFL
Theoretical investigation of nuclear excitation processes, including direct excitation through photon absorption, free electron (NEEC) and muon capture (NEμC), and Coulomb excitation (inelastic scattering). Experimental design of a table-top setup aiming at the observation of nuclear excitations in laser-plasma scenarios.

June 2019–Jan. 2023 **Charge Dynamics Electron Microscopy** EPFL
Imaging of the spatiotemporal evolution of emitted electron clouds from metallic structures upon laser irradiation using an ultrafast transmission electron microscope (UTEM). Imaging the femtosecond dynamics of charge carriers in solids through the interaction between terahertz (THz) near fields and free-electron pulses in UTEM.

June 2021–April 2022 **Imaging and control of magnetic Skyrmion** EPFL
Accomplished coherent control of skyrmion crystal rotation in discrete steps at substantially higher speeds compared to prior observations, using circularly polarized femtosecond laser pulses. I contributed to this project by developing thermal simulations to determine sample heating upon laser irradiation on a picosecond timescale using Comsol.

June 2019–Nov. 2020 **Metal-insulator transition in VO_2 using photon-induced near-field electron microscopy (PINEM)** EPFL
Employed two-color near-field ultrafast electron microscopy to visualize the photo-induced insulator-to-metal transition in an individual VO2 nanowire through the related electronic dynamics.

Publications and Conferences

Peer-Reviewed Publications

1. **S. Gargiulo**[†], I. Madan, and F. Carbone. Nuclear Excitation by Electron Capture in Excited Ions, in *Physical Review Letters*, vol. 128, n. 21, 212502, 2022.
2. **S. Gargiulo**[†], M. F. Gu, F. Carbone, I. Madan. Nuclear Excitation by Free Muon Capture, in *Physical Review Letters*, vol. 129, n. 14, 142501, 2022.
3. Y. Wu, **S. Gargiulo**, F. Carbone, C. H. Keitel, A. Pálffy. Dynamical control of nuclear isomer depletion via electron vortex beams, in *Physical Review Letters*, vol. 128, n. 16, 162501, 2022.
4. I. Madan[⋆], Eduardo JC Dias[⋆], **S. Gargiulo**[⋆], F. Barantani[⋆], M. Yannai, G. Berruto, T. LaGrange, L. Piazza, T. T. Lummen, R. Dahan, I. Kaminer, G. M. Vanacore, F. J. García de Abajo, and Fabrizio Carbone. Charge dynamics electron microscopy: nanoscale imaging of femtosecond plasma dynamics, in *ACS Nano*, vol. 17, n. 4, p. 3657–3665, 2023.
5. X. Fu[⋆], F. Barantani[⋆], **S. Gargiulo**[⋆], I. Madan, G. Berruto, T. LaGrange, L. Jin, J. Wu, G. M. Vanacore, and F. Carbone. Nanoscale-femtosecond dielectric response of Mott insulators captured by two-color near-field ultrafast electron microscopy, in *Nature Communications*, vol. 11, n. 1, p. 1–11, 2020.
6. I. Madan, G. M. Vanacore, **S. Gargiulo**, T. LaGrange, and F. Carbone. The quantum future of microscopy: Wave function engineering of electrons, ions, and nuclei, in *Applied Physics Letters*, vol. 116, n. 23, 230502, 2020.
7. P. Tengdin, B. Truc, A. Sapozhnik, L. Kong, N. del Ser, **S. Gargiulo**, I. Madan, T. Schönenberger, P. R. Baral, P. Che, A. Magrez, D. Grundler, H. M. Rønnow, T. Lagrange, J. Zang, A. Rosch, and F. Carbone. Imaging the ultrafast coherent control of a skyrmion crystal, in *Physical Review X*, vol. 12, n. 4, 041030, 2022.
8. M. Yannai, R. Dahan, A. Gorlach, Y. Adiv, K. Wang, I. Madan, **S. Gargiulo**, F. Barantani, E. J.C. Dias, G. M. Vanacore, N. Rivera, F. Carbone, F. J. García de Abajo, and I. Kaminer. Ultrafast Electron Microscopy of Nanoscale Charge Dynamics in Semiconductors, in *ACS Nano*, vol. 17, n. 4, p. 3645–3656, 2023.
9. I. Madan, V. Leccese, A. Mazur, F. Barantani, T. LaGrange, A. Sapozhnik, P. M. Tengdin, **S. Gargiulo**, E. Rotunno, J.-C. Olaya, I. Kaminer, V. Grillo, F. Javier García de Abajo, F. Carbone, and G. M. Vanacore. Ultrafast transverse modulation of free electrons by interaction with shaped optical fields, in *ACS photonics*, vol. 9, n. 10, p. 3215–3224, 2022.
10. E. Rotunno[⋆], **S. Gargiulo**[⋆], G. M. Vanacore, C. Mechel, A. Tavabi, R. E. Dunin Borkowski, F. Carbone, I. Madan, M. Zanfrognini, S. Frabboni, T. Guner, E. Karimi, I. Kaminer, and V. Grillo. One-Dimensional "Ghost Imaging" in Electron Microscopy of Inelastically Scattered Electrons, in *ACS photonics*, vol. 10, n. 6, p. 1708–1715, 2023.
11. A. A. Sapozhnik, B. Truc, P. Tengdin, E. Viñas Boström, T. Schönenberger, **S. Gargiulo**, I. Madan, T. LaGrange, A. Magrez, C. Verdozzi, A. Rubio, H. M.

Rønnow, and F. Carbone. Light-Induced Metastable Hidden Skyrmion Phase in the Mott Insulator Cu_2OSeO_3, in *Advanced Materials*, 2023.
12. E. J. C. Dias, I. Madan, **S. Gargiulo**, F. Barantani, M. Yannai, G. M. Vanacore, I. Kaminer, F. Carbone, and F. J. García de Abajo. Generation and control of localized terahertz fields in photoemitted electron plasmas, in *Nanoscale Advances*, 2023.

Articles Under Review

13. K. Akbari, **S. Gargiulo**, F. Carbone, and F. J. García de Abajo. Coherent Smith-Purcell γ-Ray Emission, *arXiv:2203.05990.*
14. **S. Gargiulo**[†], I. Madan, B. Truc, P. Usai, K. Beeks, V. Leccese, G. M. Vanacore, and F. Carbone. Revisiting the excitation of the low-lying ^{181m}Ta isomer in optical laser-generated plasma.*Under Review in Physical Review Letters*.
15. V. Leccese, M. Caldara, S. Bisi, M. Pagano, **S. Gargiulo**, C. Trigila, A. Bertsch, A. Mapelli, F. Carbone. Microstructured plastic scintillators for high resolution and high frame rate ion beam profiling.

⋆ Equally contributing authors. † Corresponding authorship.

Seminars

- Contributed talk at the 75th SUSSP and 20th STFC Summer School in Nuclear Physics and its Applications, Scotland, UK, August 2019.
- Contributed talk at the CLEO Conference, San Jose, USA [virtual], May 2020.
- Contributed talk at the DNP Fall Meeting, Boston, USA [virtual], October 2021.
- Contributed talk at the APS April Meeting, New York, USA, April 2022.
- Contributed talk at the MRS Spring Meeting, Honolulu, USA, May 2022.
- Invited talk at the Army Research Laboratory, Adelphi, USA, April 2022.
- Invited talk at TRIUMF, Vancouver, Canada [virtual], July 2022.
- Plenary invited talk at the Joint Annual Meeting of SPS and ÖPG, Basel, Switzerland, September 2023.

Posters (selected)

- "Stimulating nuclear excitation in laser-plasma scenario", Time domain control of atomic shell for nuclear excitation, Lerici, Italy, October 2019.
- "Nuclear Excitations by Electron Capture in Excited Ions", MUST Annual Meeting, Grindelwald, CH, August 2021.
- "Nuclear Excitations by Electron Capture in Excited Ions", Joint Annual Meeting of ÖPG and the Swiss Physical Society (SPS), Innsbruck, Austria, September 2021.

- "Nuclear Excitations by Electron Capture in Excited Ions", 100 Years of Nuclear Isomers, Berlin, Germany, May 2022.
- "Nuclear Excitations by Free Muon Capture", LISA Winter School—Structure of Complex Atoms, Bagnoles de l'Orne, France, September 2022.
- "Electromagnetic and atomic processes for nuclear isomer excitation in optical laser-generated plasma ", Atomic Processes in Plasma held at the International Atomic Energy Agency (IAEA), United Nations, Vienna, Austria, May 2023.

Schools

- 75th SUSSP and 20th STFC Summer School in Nuclear Physics and its Applications, Scotland, UK, August 2019.
- Quantum Computing Hard- and Software Summer School, Lausanne, CH, June 2022.
- Structure of Complex Atoms, Bagnoles de l'Orne, France, September 2022.

Other Conferences Attended (selected)

- International Online Workshop "Electrons, Photons and Plasmons-2021", March 2021, virtual.
- 3rd International Conference on Nuclear Photonics (NP2020), June 2021, virtual.

Events on Cross-Disciplinary Topics

- 72h VersusVirus Hackathon, top 6—Economics and Policy https://devpost.com/simonegargiulo2**Info** (link).
- Global Young Scientists Summit (GYSS) 2023, Singapore, January 2023 [virtual].
- Innosuisse Lausanne Business Concept, Lausanne, CH, February-June 2023.
- London Critical Theory Summer School—Social Philosophy, Birkbeck University, June 2023.

Peer-Review Activities

- Physical Review Letters

Public Outreach

Designed and produced multimedia content, including illustrations, video editing, and animations, to successfully convey the nuclear physics projects carried out by the LUMES group at EPFL. Spearheaded the creation of captivating comic art, images, and informative video, all of which were showcased on the LUMES group's official webpage:

- Google funds EPFL research on nuclear phenomena—**Cover Image** (https://actu.epfl.ch/news/google-funds-epfl-research-on-nuclear-phenomena/).
- The quantum future of electron microscopy—**Comic Art** (https://www.epfl.ch/labs/lumes/perspective-article-on-the-future-of-quantum-microscopy-wave-function-engineering-of-electrons-ions-and-nuclei/).
- Nuclear Excitation by Free Muon Capture—**Video** (https://www.youtube.com/watch?v=ps46Yb__Des).

Created a BLENDER animation illustrating the interaction of a laser pulse with a metallic structure, leading to the generation of a photoemitted electron cloud.

- Charge Dynamics Electron Microscopy—**Cover Image** (https://pubs.acs.org/doi/10.1021/acsnano.2c10482).

Teaching

Teaching Assistant for Courses

- Physique Générale (General Physics), 1st year B.Sc. Géosciences, UNIL.
  Lecturers: Fabrizio Carbone.
  Language: French.
- General Physics I, 1st year B.Sc. Physics, EPFL.
  Lecturers: Suliana Manley.
  Language: English.

Awards and Honors

- Merit Scholarship—University of Naples Federico II, 2011.
- Merit Scholarship—University of Naples Federico II, 2012.
- American Physical Society (APS) Travel Award (April Meeting 2022), Apr. 2022.
- Special Prize from Hübner & Kennedy Foundation, Sep. 2022.
- Swiss Physical Society (SPS) Award 2023 in all Physics Domains, sponsored by ABB, Sep. 2023. **EPFL news** (https://www.sps.ch/fileadmin/articles-pdf/2023/SPG-Mitteilungen_Award-Winners-2023.pdf).
- **EPFL Doctoral Program (Physics) Thesis Distinction** (Best 8%), Dec 2023. (https://www.epfl.ch/education/phd/edpy-physics/edpy-awards-laureates/).

Certifications

- NI LabVIEW CORE 1—National Instrument LabVIEW Core 1.
- COMSOL—Geometry, CAD import and Meshing.
- COMSOL—Solvers and Solution Strategies.
- AgilePM Foundation—Project Management.

Languages

Neapolitan:	native language.	**Italian**:	bilingual.
English:	full professional proficiency.	**French**:	intermediate level.

GPSR Compliance

The European Union's (EU) General Product Safety Regulation (GPSR) is a set of rules that requires consumer products to be safe and our obligations to ensure this.

If you have any concerns about our products, you can contact us on ProductSafety@springernature.com

In case Publisher is established outside the EU, the EU authorized representative is:

Springer Nature Customer Service Center GmbH
Europaplatz 3
69115 Heidelberg, Germany

The manufacturer's authorised representative in the EU is Springer
Nature Customer Service Centre GmbH, Europaplatz 3, 69115 Heidelberg,
Germany. If you have any concerns regarding our products, please
contact ProductSafety@springernature.com

Printed and bound by CPI Group (UK) Ltd, Croydon, CR0 4YY
01/10/2026
02225919-0007